D1605740

PHYSICS FOR GEOLOGISTS

PHYSICS FOR GEOLOGISTS

Richard E. Chapman

First published in 1995 by UCL Press

UCL Press Limited
University College London
Gower Street
London WC1E 6BT

British Library Cataloguing-in-Publication Data
A CIP catalogue record for this book is available from the British Library.

ISBNs: 1-85728-259-0 HB
1-85728-260-4 PB

Library of Congress Cataloging-in-Publication Data
Chapman, Richard E.
Physics for geologists / Richard E. Chapman.
p. cm.
Includes bibliographical references and index.
ISBN 1-85728-259-0. — ISBN 1-85728-260-4 (pbk.)
1. Geology. 2. Physics. I. Title.
QE28.C374 1994
550'.1'53—dc20

94-32051
CIP

Typeset in Times New Roman.
Printed and bound by
Biddles Ltd, Guildford and King's Lynn, England.

CONTENTS

This book is dedicated,
as its predecessors have been, to
my long-suffering and patient
companion of 39 years,
my wife, June,
with love and affection.

PREFACE

Geology is the study of the Earth in all its aspects, except those that are now considered to be separate sciences of the Earth, such as geophysics and meteorology. It concerns the materials of which the Earth is made, and the processes that operate on them. Very many of these processes are physical, and their understanding involves an understanding of the underlying physics. The raindrop that falls and makes its contribution to erosion is first created by condensation, falls under the influence of gravity, is held together by surface tension, reaches its terminal velocity as a result of friction long before hitting the Earth, has potential energy during its fall, and kinetic energy that is converted to mechanical energy and work when it strikes the Earth.

If sufficient drops fall, they may coalesce at the surface and begin to flow – gravity and friction are again involved, but the friction is in two forms: viscosity of the water, and external frictional resistance to flow. If the water coalesces in such a way that it infiltrates the soil or rock, then the same physical aspects are involved in its flow, but the laws are more complex and the flow of two fluids (water in one direction, air in the other) may well be involved.

When a volcano erupts, there is an obvious demonstration of energy, from the more docile outpouring of molten lava to the more explosive eruptions that distribute solid and liquid material over a wide area, and very fine material to a great height. Its internal plumbing is also governed by the laws of fluid mechanics. The emplacement of salt domes, as in Louisiana and Iran and many other parts of the world, can be modelled with immiscible fluids of smaller viscosity, and appears to be governed by the laws of fluid dynamics.

The work of the seas and oceans has a profound effect on our coastlines, and on sediments and their geological record. Around our coasts we see beaches made up of the debris created by the energy of the wind and waves. Surfboard riders know how waves are refracted around promontories; sailors know how the wind and the waves are usually related in direction, and that the strength of the wind affects the size of the waves; swimmers know that waves have energy and can stir up the sand on the sea floor, even quite coarse sand; scuba divers know that the energy of the waves diminishes with depth. Indeed, the scuba diver at oceanic margins will be unhappy with the definition of *wave base* in two American Geological Institute publications:[1] "*The depth at which wave action no longer stirs the sediments; it is usually about 10 meters*". Sediment on continental shelves can be stirred by ocean *swell* at much greater depths than 10 m – at least to 100 m – as we shall see. The definition above would only be true of waves of short length up to about 7 m or 8 m.

Physics is also important in the definition of many terms, particularly in groundwater studies, and these will be examined closely. Some of the definitions found in glossaries and dictionaries are inconsistent with the physics of the phenomenon in question.

The physics involved in geology is not confined to the scale of our own perceptions. The climatic regions of the Earth are very largely determined by the circulation of the atmosphere between the warm Tropics over a large area and the cold poles over a small area. Convection is a fairly straightforward topic of meteorology, but is far more difficult in the context of the interior of the Earth. The mid-ocean ridges and the plates seem to require convection within the Earth, but a full-scale plausible convection process has not yet been developed and remains speculative to some extent at least. At the other end of the scale, the so-called absolute age determination of rocks involves the decay of radioactive elements and some understanding of particle physics.

Geologists cannot ignore the extraterrestrial influences. The seasons are the result of our attitude in our orbit – the fact that our axes of rotation, about the Sun and about the Poles, do not coincide. Tides, which can give rise to strong currents over large areas near land, are caused by the interaction of the Sun and the Moon on the oceans. Meteorites light up our night skies by the incandescence caused by atmospheric friction, and those that survive this to reach the Earth show signs of this heat.

The principle of uniformitarianism insists that the *processes* were essentially the same in the past as they are today. The seas, tides, waves, wind, rain, rivers, all shaped the Earth in the past, as did earthquakes and volcanic eruptions.

Data acquisition is an important aspect of geology, and we must frequently do this indirectly by physical means in the interests of economy. For example, we need to distinguish between natural gas, crude oil and water in the rocks. This can most readily be achieved by measuring the resistivity or conductivity of the rock, or its response to radioactive bombardment. Shales and mudrocks can be distinguished (usually) from other lithologies by their natural γ radioactivity. We need to know the porosity of rocks that contain fresh water, or crude oil or natural gas, and this can be obtained by measuring the speed of sound through the rock, because the larger the pore volume the slower the speed of sound through the rock. The analysis of data using mathematical statistics is also an important part of science, made easy by electronic computers and calculators.

1. R. L. Bates & J. A. Jackson (eds), *Dictionary of geological terms* (3rd edn, New York: Doubleday, 1984); and R. L. Bates & J. A. Jackson (eds), *Glossary of geology* (3rd edn, Alexandria, Virginia: American Geological Institute, 1987).

Structural geology is very much concerned with the mechanical effects of forces acting on rocks, geophysics with the passage of elastic waves through the rocks, oceanography with tides and waves, hydrology with the flow of fluids on and through rocks, and meteorology with the circulations of the atmosphere. These are all specialities that will attract students, but knowledge of one field should not be regarded as excusing one from a nodding acquaintance with the others. All students will learn much about structural geology and the principles of geophysics; few will learn much about the others during their undergraduate years. Nevertheless, we must learn the things we need to know, and we need to know how waves work, how water flows, why tides vary from day to day and season to season, and many other things.

There are two problems for the student of geology: the principles of the physics involved, and the mathematics of dealing with them. The principles are much more important than the mathematics, but the mathematics is not often beyond high-school level. The mathematics tends to be either simple or intractable. For example, the flow of water through a sandstone can be treated with high-school maths or very sophisticated maths. It is most unlikely that the sophisticated actually advances our understanding of fluid flow in a practical sense. In this regard, the annual report of the US Geological Survey for 1897–8 (published in 1899) is interesting. It has an article by F. H. King on groundwater, totally devoid of mathematics, that can be read today with profit. It also has a mathematical paper on groundwater movement by C. S. Slichter that should be read with care (if at all) because it is erroneous in physical principle. If you need to brush up your mathematics, Rosen's *Mathematics recovered for the natural and medical sciences* (London: Chapman & Hall, 1992) is strongly recommended. There is also a useful summary of the mathematical background to applied geophysics in *Applied geophysics* by Telford et al. (Cambridge: Cambridge University Press, 1990; 726–44).

This book is a personal view, with emphasis on those things that I have found useful or interesting, and the things I have discovered for myself in the effort to understand geology. In particular, it is my experience that appreciation of the *dimensions* of quantities is essential for their proper understanding by the non-physicist. There are two viscosities and two permeabilities, and they differ in their dimensions. So we start with dimensions and related topics because there is constant reference to them in the text. If the reader does not feel these needs, then this part can be omitted.

It is not my intention to turn a geologist into a geophysicist – far from it. Geophysicists (and, indeed, hydrologists, meteorologists and almost all the specialities) require much greater depth to their studies than this book provides. This book is intended as a help to students of general geology, to those who thought they would not need their school physics any more and to those who did less than they now need. The purpose is to improve their understanding of geology by improving the understanding of the physics involved in geology and geological processes. If you find any errors, inaccuracies, or unclear passages, please write to the author. That would be a valuable service for the author and, perhaps, future students.

The contents are not exclusively directed to aspects of geology. Some topics come under the heading of general knowledge of the Earth and Universe for physical scientists. Kepler's and Newton's laws of motion come into this category perhaps, but it is a very good starting point for understanding force, not just for understanding the orbits of planets and their satellites. Momentum is important when considering the effects of a collision between the Earth and another celestial object.

Readers will notice that many of the scientific principles were established three or four centuries ago, and that some names reappear in different contexts. Although Newtonian physics may not satisfy a physicist, it is entirely satisfactory for geologists and geology when dealing with all but particles. It is very rare nowadays for any scientist to make a mark in more than one field, but science today requires the broadest possible understanding of scientific disciplines if we are to solve the problems that cannot be confined to one.

ACKNOWLEDGEMENTS

I make no claim to originality in this book. Inevitably, in a work such as this, I have relied largely on the work of others; but to have acknowledged this at every step would have made it unreadable. I therefore acknowledge here with particular gratitude the help I have received from the following works:

- *The new Encyclopædia Britannica* (15th edn, Chicago: Encyclopedia Britannica, 1984) – many articles consulted over the years apart from those cited.
- Feynman, Leighton & Sands, *The Feynman lectures on physics.* (Reading, Mass.: Addison–Wesley, 1963–5).

However, it is not all the work of others. The sections on sliding, fluid flow, and settling velocities of solids in fluids, are based on my work, which will be found in the references. I owe, as always, a debt of gratitude to the late M. King Hubbert, whose early papers on fluid flow and mechanics in geology, inspired me.

I am also grateful to the late Dr J. P. Webb, friend, colleague and geophysicist at the University of Queensland for many years, who helped me with many points in the early stages of this book (as he had with earlier books); to Drs K. Whaler and Stan Murrell for very helpful comments and suggestions on two drafts. Many of my former colleagues were consulted on various points, and I am grateful to them too. For the positions of the magnetic poles I am indebted to C. E. Barton of the Australian Geological Survey Organization, formerly the Bureau of Mineral Resources, Geology & Geophysics, Canberra. All remaining mistakes are my own.

Glaston Hill
Upper Camp Mountain Road
Camp Mountain, Australia 4520

SYMBOLS

Greek

$\alpha, \beta, \theta, \phi, \varphi$ angles
γ specific heat capacity ratio, weight density
Δ dilatation
Δt sonic/acoustic transit time
ε extraordinary ray, strain
η coefficient of viscosity
λ Lamés parameter, wavelength
μ coefficient of friction, coefficient of reflection, permeability of free space
ν frequency, kinematic viscosity
Φ fluid potential
Ω ohm
ρ mass density
σ stress, surface tension
τ shear stress
ϕ radiant flux
ω angular velocity, ordinary ray

Roman

a acceleration
c speed, velocity
E energy
F force
f porosity (fractional)
G gravitational constant, Lamé's parameter, modulus of rigidity
g acceleration due to gravity
h total head
K hydraulic conductivity, bulk modulus
k intrinsic permeability
I luminous intensity
m mass
n refractive index
p pressure
R gas constant, hydraulic radius, reflectance
r correlation coefficient, radius
S vertical component of total stress
T period ($1/\nu$), temperature
V speed, velocity, volume
v Poisson's ratio
z depth

1 BASIC CONCEPTS

Those of us who are not professional physicists, and those who last did some physics many years ago, may have forgotten some of the things that physicists take for granted. We start, therefore, with most of the basic concepts that will be used throughout this book.

Dimensions and units

When we write an equation, such as that relating pressure to vertical depth below the surface of a body of static water,

$$p = \rho g z \tag{1.1}$$

where ρ is the mass density of the water (see p. 129 for the Greek alphabet), g is the acceleration due to gravity (or the acceleration of free fall) and z is the depth below the free surface, two aspects are involved: the *dimensions* of the constituent parts, and the *units* we are to use. It is not at all clear that multiplying a density by an acceleration and by a length has any real meaning; but if we use the "right" units, we can evaluate the equation and determine the pressure at any depth in any liquid.

All valid equations must be *dimensionally homogeneous*, that is, the dimensions must be the same on each side of the equation, and true constants are dimensionless. Units must be consistent for reliable evaluation. Inconsistent units are those in which components vary from one unit to another, such as pounds per square inch (psi) with lengths measured in feet. The requirement of dimensional homogeneity is used in *dimensional analysis* in order to find the form of an equation when all the quantities involved can be listed. This could be a powerful tool in Earth Sciences in which analytical solutions to problems are not always possible (or are beyond our abilities).

Many physical quantities have one or more of the dimensions of mass, length, and time, or are dimensionless. An angle is dimensionless, being the ratio of a length to a length. Dimensions are indicated by the initial capital or lower case letter with its exponent (if other than 1). So an area has the dimensions of a length multiplied by a length, or L^2; a velocity, a length divided by time, or LT^{-1}. Dimensions may be put in square brackets after a quantity, e.g. mass density [ML^{-3}], and for a dimensionless quantity [0], or written out. It is essential for clear thinking in science to consider the dimensions of quantities. Table 1.1 lists the common quantities. Dimensions are quantified using arbitrary units such as metres for length.

Table 1.1 Dimensions of some physical quantities.

acceleration	LT^{-2}
bulk (and other) moduli	$ML^{-1}T^{-2}$
compressibility	$M^{-1}LT^{2}$
density, mass	ML^{-3}
density, weight	$ML^{-2}T^{-2}$
energy	$ML^{2}T^{-2}$
force	MLT^{-2}
frequency	T^{-1}
hydraulic conductivity	LT^{-1}
inertia	M
moment of inertia	ML^{2}
torque	$ML^{2}T^{-2}$
momentum	MLT^{-1}
permeability, coefficient of (fluid)	LT^{-1}
permeability, intrinsic (fluid)	L^{2}
potential (fluid)	$L^{2}T^{-2}$
pressure, stress	$ML^{-1}T^{-2}$
specific discharge	LT^{-1}
surface tension	MT^{-2}
velocity	LT^{-1}
viscosity, absolute or dynamic	$ML^{-1}T^{-1}$
viscosity, kinematic	$L^{2}T^{-1}$
weight	MLT^{-2}
work ,	$ML^{2}T^{-2}$
temperature	θ
quantity of heat	$ML^{2}T^{-2}$
thermal conductivity	$MLT^{-3}\theta^{-1}$

To be valid, Equation 1 must also balance dimensionally, that is, the sum of the exponents of M, the sum of the exponents of L, and the sum of the exponents of T must be equal on both sides of the equation. Pressure is a force (mass times acceleration) on an area and therefore has dimensions $MLT^{-2}/L^2 = ML^{-1}T^{-2}$; mass density is mass per unit of volume and has the dimensions ML^{-3}; acceleration has the dimensions LT^{-2}; depth has the dimension of length. So, Equation 1, written as dimensions, is

$$\begin{array}{ccccccc} ML^{-1}T^{-2} & = & ML^{-3} & LT^{-2} & L & = & ML^{-1}T^{-2} \\ p & & \rho & g & z & & \end{array}$$

and it is seen to balance. We shall examine this in more detail shortly.

To evaluate equations such as Equation 1 we need to decide on the *units* of measurement. These are clearly arbitrary, and the earliest measures were those readily to hand (like the inch, which was the length of a thumb), or to foot. In earlier times, the *size* of a quantity determined the unit chosen. So we had inches, feet, yards and miles for length; and there were 12 inches in a foot, 3 feet in a yard and 1760 yards (5280 feet) in a mile. If pressures are measured in *pounds per square inch* (psi) and lengths in *feet*, equations such as Equation1 require a factor to take this into account.

The *cgs system* (**c**entimetre, **g**ram[1], **s**econd) was also in general use, but one had to remember to use centimetres even for very large lengths. It was also called the metric system. Most countries nowadays use the MKSA system, with basic units of metre, kilogram, second, and amp; or SI units (*Système International d'Unités* usually just written SI), which is based on the MKSA system and will be discussed below.

We make a distinction between fundamental or *basic* quantities, and *derived* quantities. The common fundamental dimensions are those of *mass*, *length*, and *time*. The common derived quantities include *speed* or *velocity* $[L\,T^{-1}]$, *acceleration* $[L\,T^{-2}]$, *energy* $[M\,L^2\,T^{-2}]$, *pressure* $[M\,L^{-1}\,T^{-2}]$, and *force* or *weight* $[M\,L\,T^{-2}]$. Table 1.1 gives the common quantities and their dimensions. From this you can see that Einstein's famous equation involving energy, mass and the speed of light, $E = mc^2$, satisfies the requirement of dimensional homogeneity except that the nature of the speed is not identified from the dimensions alone. Energy has the dimensions $M\,L^2\,T^{-2}$, so the quantities postulated for the right-hand side must also have these dimensions with the same exponents. If mass is one of these, the other could be $L^2\,T^{-2}$, which is a speed squared, or a length multiplied by an acceleration. There may be more than one plausible solution (and if you can find one for Einstein's equation, your future is assured). True constants, being dimensionless, do not come out of the analysis but must be determined experimentally. Constants that are not dimensionless are called material constants. You may ask, is not the speed of light, c, a constant? The answer is yes and no. The speed of light varies in different media but is considered to be a constant in a vacuum. It has dimensions and so the *number* depends on the units used. A true constant in a mathematical sense is independent of the units used (provided they are consistent).

SI units

Science uses SI units (*Système International d'Unités*, so spelled, not *Internationale*). SI units are entirely consistent. Table 1.2 lists the common *basic* units, that

1. The spelling, gramme and kilogramme, is permitted; but Australia, UK and USA have adopted the shorter spelling.

is, ones that form the building blocks of science. Table 1.3 lists the common named derived units. There is some confusion in the USA, because their industries, and even some of their scientific journals, continue to use units that are not consistent. Twenty years ago, 136 countries were metric or changing to metric and the only countries not going metric were Barbados, Jamaica, Liberia, Nauru, Sierra Leone, S. Yemen, Sultanate of Muscat & Oman, Tonga and the USA!

Table 1.2 Basic SI units.

mass	kg	kilogram
length	m	metre
time	s	second
electric current	A	ampere
temperature	K	kelvin
amount of substance	mol	mole
luminous intensity	cd	candela
plane angle	rad	radian
solid angle	sr	steradian

Table 1.3 Common named derived SI units.

Unit name	Abbreviation	Nature	Units
hertz	Hz	frequency	s^{-1}
newton	N	force	$kg\,m\,s^{-2}$
joule	J	work, energy, heat	N m
pascal	Pa	pressure	$N\,m^{-2}$
watt	W	power	$J\,s^{-1}$
volt	V	electric potential diff.	$W\,A^{-1}$
ohm	Ω	resistance	$V\,A^{-1}$
siemens	S	conductance	Ω^{-1}
coulomb	C	electric charge	A s
henry	H	inductance	$V\,A^{-1}\,s$
tesla	T	magnetic flux density	$V\,s\,m^{-2}$
weber	Wb	magnetic flux	V s
lux	lx	illumination	$cd\,sr\,m^{-2}$

The degree *Celsius*, °C, has the same value as a kelvin, but the scale is measured from the freezing point of pure water (273·15 K). Centigrade is not the *SI* unit and should not be used. For the measurement of angles the "decimal" degree, the *grade*, with 400 degrees in a circle, never received acceptance and we continue with 360° to the circle, each degree being divided into 60 minutes (′) and each minute, into 60 seconds (″). But the radian is more fundamental, being the angle subtended at the centre of a circle by an arc of the circumference that is equal to its radius. So there are 2π radians in a circle of 360°, and a radian is 57·3° (very nearly). Computers use radians, but it is not a convenient measure for the dip of a rock, for instance, 10° being 0·1745 radians. A *nautical mile* (1852 m) is retained for navigation because it is a minute of arc of latitude (its length changes slightly with latitude because of the flattening of the Earth, and its length on a Mercator chart varies with latitude due to the projection) and the *knot* because it is a nautical mile per hour. The *litre* was redefined in 1964 as 10^{-3} m^3. This should not be used for precise measurements because of possible ambiguity. Between 1901 and 1964 the litre was defined as the volume of one kilogram of pure water at 4°C and one standard atmosphere (the pressure exerted by 760 mm of mercury at 0°C, or 101·325 kPa) and this volume was $1{\cdot}000\,028 \times 10^{-3}$ m^3.

Abbreviations The abbreviations of SI units are rigid and invariable (Tables 1.2, 1.3, 1.4). They stand by themselves, singular and plural, without a stop (unless it is the end of a sentence). It is unfortunate that some countries adopted them carelessly, so you may see advertisements that it is *6 Klms* or *6 K* or *250 Mtrs* to wherever: it should be just *6 km* or *250 m*.

Table 1.4 Prefixes for SI units.

10^{-18}	atto	a
10^{-15}	femto	f
10^{-12}	pico	p
10^{-9}	nano	n
10^{-6}	micro	μ
10^{-3}	milli	m
10^{-2}	*centi*	c
10^{-1}	*deci*	d
10	*deka*	da
10^{2}	*hecto*	h
10^{3}	kilo	k
10^{6}	mega	M
10^{9}	giga	G
10^{12}	tera	T

The use of those units in *italics* is discouraged.

For those units named after a person, the abbreviation is written as a capital, but the name of the unit without – except Celsius, °C. So *pascal* (Pa) and *siemens* (S), the latter retaining its final s in the singular because that is the name.

For very large and very small numbers of units there is a set of prefixes (Table 1.4). The basic unit *kilogram* already has a prefix and it does not take another. A thousand kg is one *tonne* (t), a unit of *mass*; or Mg. The use of those in *italics* in Table 1.4 is discouraged, but *centimetre* (cm), *hectare* (ha, 10 000 m^2) and *hectopascal* (hPa, meteorology) will no doubt persist.

When these prefixes are used in combination with a basic or derived unit, any exponent applies to both. So 1 km^2 is 10^6 m^2 and 1 mm^2 is 10^{-12} m^2. Numbers should be written with a space, not a comma, separating groups of three digits on either side of the decimal point, so 12 345·678 9; and with a space between the number and any units (except perhaps the degree symbol, °). Four-digit numbers may be written together, as 1234 and 1234·5678. A number between 1 and –1 must always have a zero before the decimal point: 0·75, not ·75. The decimal point should be returned to mid-line, · ; only the limitations of the manual typewriter led to its being combined with the full stop or period.

There are advantages, arising from the prefixes, in using what is called the engineering notation of numbers, that is, giving the exponent of 10 to a multiple of three (the practice used here), rather than the scientific notation. So 10^{10} in scientific notation will be written 10×10^9.

Distinctions and definitions

Average, mean, median

The word *average* has several meanings in shipping and science, but it usually means the *arithmetic mean*, $\Sigma x/n$, where Σx is the sum of n values of x. It follows that the word *mean* is at least partly synonymous with "average". The mean is usually written with a bar over the symbol, e.g. x and $\bar{\rho}$.

The *weighted average* is the sum of the products of the quantity and the weight, w, attached to it, divided by the sum of the weights, $\Sigma xw/\Sigma w$. This is used, for example, in sedimentology for representing the results of a sieve analysis, where w is the proportion of the sediment by weight retained on sieve sized x.

The *geometric mean* of n positive quantities is the nth root of the product of these quantities: $(x_1 \times x_2 \times x_3 \times \ldots \times x_n)^{1/n}$. This is used in sedimentology because size distribution of sediments is normalized by its use, becoming more symmetrical when plotted as a frequency diagram. A linear size scale would put almost all sediment into the range 10^{-6} to 10^{-3} m, creating difficulties for the larger size ranges.

The *harmonic mean* is the reciprocal of the arithmetic mean of the reciprocals of the values to be averaged: $n/\Sigma(1/x)$. This could have its uses in sedimentology because the harmonic mean grain size of a sediment is better related to its permeability than other *measures of central tendency*, as these quantities are sometimes called.

The *median* is that value of a set of values that divides the set into two halves. If your class lined up in order of increasing height (or weight, or wealth, or any other measurable quantity), the median would belong to the student or pair of students with the same number of students on either side. In a sediment, it is the grain size that has equal *weight* larger and smaller. Life is too short to count grains!

Averages do not always have useful meanings in science. Limestones consist primarily of calcite, so CaO and CO_2 dominate its composition. MgO may also be present; but most limestones have either less than 4 per cent or more than 40 per cent $MgCO_3$ because substitution of Mg in the lattice occurs only with difficulty between these figures. So the average, 16·5 per cent, is most misleading.

Solids, fluids (liquids, gases)

The distinction between solids and fluids is important in science and engineering, but in Earth science it is perhaps a little more subtle than in the other sciences (for reasons that will become apparent). A fluid is defined as a substance that yields *at once* to shear or tangential stress. Engineers sometimes define fluids as material that yields *in time* to the slightest shear stress (and so solids do not yield in time to the slightest shear stress). This is satisfactory if the timescale is short; but it obscures the difference between solids and fluids for geologists because *most* materials that an engineer would regard as solid will yield to slight shear stress over very long periods of time. This is true of most materials of the Earth.

In eastern Venezuela and in Trinidad there are so-called pitch lakes – pitch that has welled up from the subsurface like a volcano. This pitch can be broken with a hammer, and so satisfies the definition of a solid in the sense that it has not behaved as a fluid. Yet it has behaved as a fluid in flowing from the subsurface, and spreading out at the surface without fracture.

Earth scientists must remain flexible about the distinction between solids and fluids because if they follow the engineer's definition, they will fail to understand geological processes; if they always insist on the geological evidence, they will lose their friends and be regarded as prematurely odd.

Scale has an effect on movement, particularly speed. The child walks with quicker steps than the parent; the swallow flaps its wings more frequently than the eagle; and you can balance a broomstick on the end of your finger easily, but not a matchstick. When making true scale models, all the relevant dimensions must be scaled in the right proportion, and if we wish to scale a process that takes millions of years so that we can see its effect in days, we use material in the model that is much more fluid, such as wax. See Hubbert (1937) for a better understanding of the theory of scale models, and Ramberg (1981) for interesting applications to geology.

Speed, velocity, and acceleration

As you drive through a town, your speedometer tells you your *speed* $[L\,T^{-1}]$ over the ground. This is variable and, knowing the distance travelled and the time taken to travel it, your average speed can be calculated. Your route will probably not have been straight, so that progress towards your goal has been slower than your progress over the road. Velocity is the net speed in a particular direction. It is a *vector*: it has magnitude and direction. Speed is a *scalar*: it has magnitude but no direction. A yacht tacking to windward may be travelling at 6 knots through the water, but its velocity to windward will be about $6/\sqrt{2} \approx 4\frac{1}{4}$ knots. If a current is flowing, the velocity over the sea floor will be different again. Writers (including this one) are not always consistent in making this distinction, but the context usually makes it clear.

Acceleration $[L\,T^{-2}]$ is the rate at which speed or velocity changes with time. It has direction and magnitude and is therefore a *vector*. Acceleration in physics is not necessarily in the direction of motion, or opposed to it. It is in the direction of the force applied that causes the acceleration. A football kicked in the air has a downward acceleration due to gravity that gives it its trajectory. The component of velocity parallel to the ground would be constant if there were no air resistance.

Frequency, wavelength, amplitude and phase

As waves come in to the coast, the distance between them – the wavelength, λ – is fairly constant but gets shorter as the beach is approached; the height of the wave from trough to crest (or amplitude, which is half the height of the wave) is fairly constant, but gets greater as the beach is approached; and the frequency (symbol ν, the Greek letter nu) or period ($1/\nu$ or T) with which the waves break on the beach remains fairly constant. The speed or velocity (c) of waves is variable and a function of wavelength. The dimensions of frequency are T^{-1}; of wavelength, L. So the relationship between velocity $[L\,T^{-1}]$, wavelength and frequency is $T^{-1} = L\,T^{-1}/L$; in other words, frequency times wavelength gives velocity or, $\lambda = c/\nu$.

Waves are additive (Fig. 1.1), so if two parallel wave trains are in the same *phase* (that is, their crests and troughs coincide) much bigger waves and deeper troughs will result. If they are exactly opposite in phase, small lumpy waves will result. And if the waves trains are not identical, not parallel, more complicated *interference* will be seen.

Dimensional analysis

One of the great advantages of considering dimensions carefully is that one can use the powerful tool of dimensional analysis. We said above that if an equation is to be valid, we must be able to use any consistent or coherent set of units and

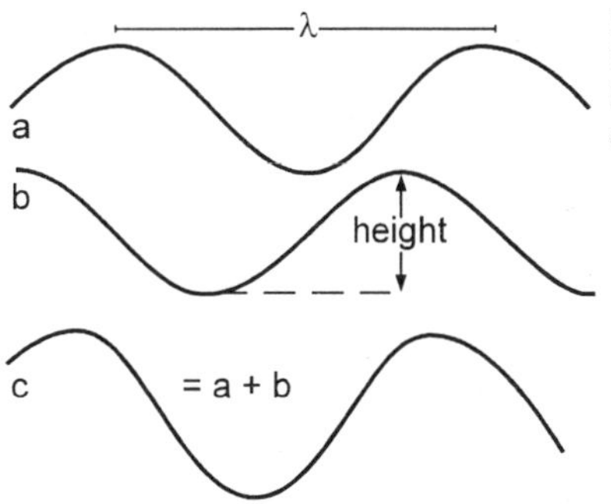

Figure 1.1 Waves are additive.
λ is the wavelength. Wave *a* is out of phase with wave *b*. Wave *c* is the wave that results when both waves *a* and *b* coexist.

any true constants must be dimensionless. It must also be dimensionally homogeneous. It is this requirement for homogeneity that allows us to analyze the dimensions with a view to deriving the form of the equation. Physical quantities such as volume, density, velocity, can be expressed in terms of one or more dimensions of **m**ass, **l**ength and **t**ime, and any equation relating such quantities must be homogeneous, and the exponents or powers of M, L, and T must be equal on both sides of the equation. For example,

weight density	=	mass density	×	acceleration due to gravity
γ		ρ		g

or, in dimensions,

$$ML^{-2}T^{-2} \quad = \quad ML^{-3} \quad \times \quad LT^{-2}.$$

To check this, we set up what are called *indicial equations*:

				=				
For M:	1			=	1			
For L:		−2		=		−3	+1	
For T:			−2	=				−2.

This is the principle of dimensional analysis: we use the dimensions of the quantities we suppose to be involved, and so seek the form of the equation relating them by equating the exponents. It must be emphasized, though, that if we omit any components, we may finish with the wrong answer. An example will make this clear.

If we did not know the equation (Eq. 1.1) relating depth to pressure, p, in a liquid, we might assume that it is a function of mass density, ρ, the acceleration due to gravity, g, and the volume, v, and write

$p = f(\rho, g, v)$ (say, pressure is a function of ρ, g, and v).

This can be written in dimensional form

$$ML^{-1}T^{-2} = (ML^{-3})^a (LT^{-2})^b (L^3)^c$$
$$= M^a L^{-3a} L^b T^{-2b} L^{3c},$$

from which we write the indicial equations

For M: $1 = a$

For L: $-1 = -3a + b + 3c$

For T: $-2 = -2b$.

The solution of these simultaneous equations is $a = 1$, $b = 1$, and substitution of these into the indicial equation for L gives $c = 1/3$. This tells us that it is not volume but length that is involved, and that length is clearly the depth at which p is required:

$$ML^{-1}T^{-2} = (ML^{-3})(LT^{-2})(L),$$

and the function is

$$p = B\rho g d,$$

where B is a dimensionless constant. A few measurements would establish that the dimensionless constant

$$B = p/\rho g d = 1$$

for a liquid of constant density.

Take another example, the results of which we will use later. The tension in a string as we twirl a stone around in a horizontal orbit, or the gravitational force that keeps a satellite in orbit, is called the *centripetal* force. What is the equation for the centripetal force required to keep an artificial satellite in a circular orbit? This force is clearly a function of the mass of the object, its speed in its orbit, and the radius of that orbit:

Force	=	f(mass,	length,	velocity).
MLT^{-2}		M^a	L^b	$(LT^{-1})^c$

The indicial equations are:

for M: $1 = a$

for L: $1 = b + c$

for T: $-2 = -c$

from which $a = 1$, $c = 2$ and $b = -1$, and we deduce that the centripetal force is given by

$$F_c \propto m V^2/r, \quad (1.2)$$

where m is the mass of the satellite, $\propto$ means "varies as" or "is proportional to", V

is its speed in its orbit, and r is the radius of its orbit; and we would find by measurement that the constant is 1 for a circular orbit.

All valid equations can be written in dimensionless terms with the great advantage in experimental work that only two variables in each have to be determined for different values, and the simplest can be chosen. According to Buckingham's Π theorem, n physical quantities expressed in d fundamental dimensions can be arranged into $(n-d)$ dimensionless groups from which the form of the equation can be derived. Each of these dimensionless groups is called a Π-term.

See Buckingham (1914, 1921), Brinkworth (1968: 79–89) or Pankhurst (1964) for a more detailed account of dimensional analysis

2
FORCE

The dimensions of force are those of a mass times an acceleration $[M L T^{-2}]$. The unit of force is the *newton*, N, which is the force that gives to a mass of 1 kg an acceleration of $1\,m\,s^{-2}$. So $1\,N = 1\,kg\,m\,s^{-2}$. Force is a quantity that has both magnitude and direction, so it is a vector.

When considering forces such as gravity acting on bodies, it is easier to consider the forces acting on the centre of mass of a body, rather than on all of the particles. That is, we consider the bodies to have mass but no size, as if its mass were concentrated at a single point. The justification for this simplification is this. If you balance a rod horizontally on your finger, the force of gravity is acting on all particles of the rod, and equally on those on either side of your finger. The effect is that of a weight equal to that of the rod on your finger, which is in the same vertical plane as what is called the *centre of gravity* of the rod – also the *centre of mass*. You could refine the experiment by using a fulcrum that ended in a sharp edge and, by rotating the rod horizontally and axially, determine with some precision the position of the centre of mass.

Force acting on a surface can be resolved into two components, one normal to the surface, the other tangential – a shear force. Forces can also be resolved by vector arithmetic into three mutually perpendicular directions, called components. When considering forces in two dimensions (which is commonly sufficient in geology), they may be resolved into

$$F_x = F \cos \beta;$$

$$F_y = F \sin \beta, \tag{2.1}$$

where β is the angle between the direction of the force and the direction into which it is resolved.

So much depends on work done more than three centuries ago, and so much can be learnt from it, that we shall start with a summary of the relevant work of Kepler and Newton.

Kepler and Newton: the laws of motion

Copernicus and Galileo played an important part in science when they interpreted the movement of the stars and planets as the Earth revolving around the Sun, and the Earth revolving about its own axis. It was Kepler and Newton, however, who put substance into these assertions.

Johannes Kepler (1571–1630) used Tycho Brahe's database (as we would now call it) of planetary positions, but first he had to solve the problem of refraction if the true position of the heavenly bodies was to be obtained from the observed position. After this had been done, he stated that:

- planetary orbits are ellipses, with the sun at one of the two foci,
- the vector radius of each planet to the Sun sweeps out equal areas of the ellipse in equal intervals of time, and
- the squares of the sidereal periods of revolution are proportional to the cubes of their mean distances. (The Earth's sidereal period of revolution is 365·26 days.)

These laws are kinematic laws; they describe motion without reference to force.

Isaac Newton (1643–1727) developed the three dynamical laws of motion:

- If a body is at rest, or moving at a constant speed in a straight line, it will continue at rest or at a constant speed in a straight line unless acted on by a force. (This is also called the Law of Inertia; it was not a new law but a restatement of Galileo's.)
- The acceleration, or rate of change of velocity with time, a, is directly proportional to the force F applied and inversely proportional to the mass, m: $a = F/m$.
- The actions of two bodies on each other are always equal and opposite (or, to each action there is an equal and opposite reaction).

The concept of force is seen to lie in its effects, and a newton is defined as the force that gives to a mass of one kilogram an acceleration of one metre per second per second.

Newton was able to derive his law of gravitation from his, and Kepler's third, laws of motion, as applied to the Moon. The argument runs like this:

- the Moon's motion is not in a straight line, therefore it is being acted on by a force
- the action of this force has an equal and opposite reaction, so the Moon is attracting the Earth and the Earth is attracting the Moon equally, and
- this attractive force gives rise to an acceleration of the Moon towards the Earth, and the Earth towards the Moon.

Newton found the kinematic relationship between the radius (r) of a *circular* orbit, its period (T), and its centripetal acceleration to be $a = 4\pi^2 r/T^2$, and from that inferred the inverse square law (see Note 1 on page 131 if you cannot do that for yourself). The Moon's period is 27·3 days, or $2{\cdot}36 \times 10^6$ s; its distance from the Earth is 384×10^6 m (sufficiently well known in Newton's day) so the Moon's acceleration towards the Earth was found to be $2{\cdot}7 \times 10^{-3}$ m s^{-2}. The ratio of the Moon's acceleration and the Earth's, $2{\cdot}7 \times 10^{-3}/9{\cdot}8$, is about 1/3600. The distance of the Moon from the Earth is about 60 Earth radii.

Newton also showed that all orbits are conic sections (that is, the trace of a plane cutting a right cone). These are circles, ellipses, parabolas or hyperbolas, depending on the angle the plane makes with the axis of the cone, and the angle of the side of the cone. Parabolic and hyperbolic orbits are never completed. (For the mathematical details, see Rosen 1992: 61 ff.)

Weight, mass, and density

The distinction between weight and mass, weight density and mass density, concerns the force of gravity. Newton's Law of Gravitation states that every particle of mass m_1 attracts every particle of mass m_2 with a force that is directly proportional to the product of their masses and inversely proportional to the square of the distance between their centres of mass (which we can take for the time being to be the centres of the objects); and this force acts along a straight line joining them:

$$F = G\,(m_1 m_2)/r^2, \qquad (2.2)$$

where r is the distance between their centres of mass, and G is the universal constant of gravitation. G is not a *true* constant because it has dimensions and the number depends on the units used:

$$[G] = [M L T^{-2}]\,[L^2]\,[M^{-2}] = M^{-1} L^3 T^{-2},$$

which are those of the force of attraction times the square of the distance between the bodies, divided by the product of their masses (N m^2 kg^{-2}). The force of attraction F between two bodies gives to each an acceleration towards the other. The amount of acceleration depends on the mass of the body because $F = m_1 a_1 = m_2 a_2$.

For objects on the surface of the Earth, there is an observable acceleration owing largely to the gravitational attraction between the centres of mass of the Earth and the object. This is denoted by g and is known as the *acceleration due to gravity* or the *acceleration of free fall*. (This is strictly a vector, having direction and magnitude; we shall come to that.) So the weight W of the object is

$$W = m\,g. \qquad (2.3)$$

The observable acceleration due to gravity, g, has several components. The largest is indeed the acceleration due to gravity, $G\,(M_E m)/R_E^2$. This is reduced everywhere on Earth except at the poles by a centrifugal acceleration due to the Earth's rotation about its axis. This is greatest at the Equator, where it amounts to about 34 mm s^{-2}, and decreases towards the poles. The total effect is more complicated than that because the equatorial bulge, caused by the centrifugal force, means that there is more mass contributing to g and a larger centrifugal force due to the larger radius; and at the poles, the surface is rather nearer the centre of mass of the Earth. These components account for almost all the value of g, but not quite all, and for weighing purposes its value may be taken as 9·8 m s^{-2}, which is close to its value at sea level in latitude 45°.

There are advantages in thinking of g not as an acceleration $[L\,T^{-2}]$ but as a weight per unit mass $[M L\,T^{-2}/M]$, not as 9·8 m s^{-2} but as 9·8 N kg^{-1}, g N kg^{-1}, because that keeps the effect of g in mind. One kilogram mass weighs g newton. Note that g is independent of the mass of the object, so Galileo was right when he postulated that all bodies would fall at the same rate in a vacuum.

Density relates to the weight or mass of unit volume, so quantitatively we must specify whether we are using weight density or mass density. The dimensions of mass density (ρ) are ML^{-3}. The dimensions of weight density (ρg or γ) are $ML^{-2}T^{-2}$. Their units are kg m^{-3} and N m^{-3}, respectively. In some practical applications (such as drilling for oil or water) weight density can be usefully considered as a pressure per unit of vertical length, Pa m^{-1}. If you are *required* to use US units, work in SI units and convert to those required. If you cannot do that in drilling practice, use psi/foot as the unit of weight density.

Bulk density, as of a rock, is the average density (weight/bulk volume, or mass/bulk volume) of a material that has two or more components of different density, such as porous rock. Wet bulk density implies water in the interstices or pore spaces; dry bulk density implies air.

Statics and dynamics

Statics is the branch of mechanics that is the study of bodies at rest and the forces that compel a body to remain at rest. Dynamics is the study of bodies in motion. Newton's laws of motion lead to the propositions that:

- bodies in motion possess inertia, requiring a force to stop or otherwise change speed or direction of the motion, and
- where forces are applied separately, the total force is the vector sum of the individual forces.

So it is a necessary condition that the sum of the forces acting on a static body be zero (hence the statement that action equals reaction).

A book resting on a table involves two obvious forces: the weight of the book and the normal reaction to that weight. The sum of these must be zero. If the table is tilted (Fig. 2.1), there are other forces, namely, the component of weight normal to the table-top, and that along the table-top. The reactions to these are the normal reaction ($W_n = W\cos\theta$), and the resistance due to friction ($W_s = W\sin\theta$). If the book is at rest, the sum of all these must be zero. As long as the frictional resistance can equal the component of weight down the slope, the book will remain stationary.

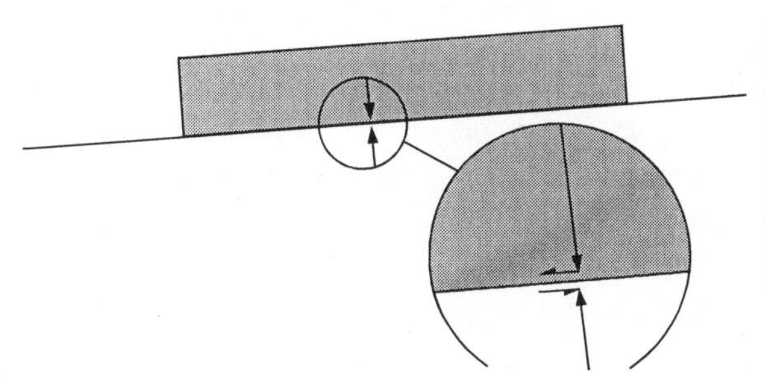

Figure 2.1 The sum of forces acting on a static body must be zero.
The normal reaction is equal to the component of weight normal to the surface; and the frictional resistance is equal to the component of weight down the surface.

When the sum of forces equals zero, the relationship can be expressed in another way called Lami's theorem, or the triangle or polygon of forces. If three forces in the same plane act at a point in a stationary object (such as the centre of mass), then three lines drawn parallel to these forces with lengths proportional to the magnitude of each force form the three sides of a triangle. So the book on an inclined surface (Fig. 2.1) is acted on by the weight of the book, the normal reaction, and the frictional resistance, and this can be represented by a triangle of forces, as in Figure 2.2.

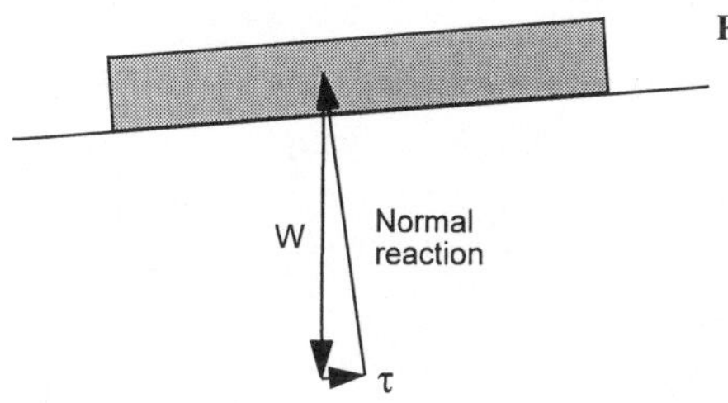

Figure 2.2 A triangle of forces.

When it slides, there is no longer a *triangle* of forces because the shear component of weight, T, is greater than the frictional resistance, τ.

If more than three forces act on a body at rest, a *polygon* of forces results.

Centrifugal and centripetal forces: motor cycles and spacecraft

When a motor cycle takes a bend in the road, it (and the rider) has to bank. If it did not bank, Newton's First Law of Motion would ensure that it went straight ahead.

A force is needed to make it change direction. Similarly, when you swing a stone around on a string, a force exists in the string that makes the stone travel in a circular orbit. Again, Newton's First Law of Motion asserts that the stone will fly off at a tangent to that circle the moment you let the string go. The force that keeps the object in its orbit is the *centripetal force* (the equal and opposite reaction being the *centrifugal force*). Dimensional analysis in the previous section revealed that this force, F_c, is equal to mV^2/r, for a circular orbit (or an acceleration of V^2/r towards the centre of the circular orbit).

When a motor cycle takes a bend of radius x m, it requires a centripetal force of mV^2/x. The machine will have to bank to provide this force[1], and bank at an angle such that the lateral component of weight of the machine and its rider provides the centripetal force required. The vertical component is the weight of the machine and rider, mg, so the lateral component is $mg \tan \theta$ and

$$mg \tan \theta = mV^2/x,$$

$$\theta = \tan^{-1} (V^2/xg). \tag{2.4}$$

Similar considerations apply to the orbits of planets around the Sun and satellites around their planets. A spacecraft in a circular orbit requires an acceleration towards the Earth that is provided by gravity, so, denoting the Earth's radius by R_E, its mass by M_E, and the gravitational constant by G (as usual),

$$V^2/(R_E + h) = G M_E/(R_E + h)^2$$

from which

$$V^2 = G M_E/(R_E + h), \tag{2.5}$$

and the relationship between a spacecraft's elevation above the surface of the Earth, h, and its required velocity can be determined. Note again that the mass of the object does not come into the equation. That is why an astronaut feels "weightless", and everything in the spacecraft, even the smallest thing, appears to be weightless. They are not weightless, because they have mass and there is mutual attraction between their masses and the Earth's. They are all accelerating towards the Earth at the same rate. This equation does not mean that no force acts on the spacecraft, for if that were true, Newton's first law of motion would come into play and the spacecraft would travel in a straight line and fly off into space. The force, the only significant force, acting on it is gravitational and centripetal, giving it an acceleration towards the Earth. It is this unbalanced force that gives rise to the forward motion and this acceleration that gives it its orbit.

1. The racing motorcyclist shifts his weight inwards on the bends so that the machine will be more upright. The angle is measured from the points of contact with the road through the centre of gravity of bike and rider as a whole.

Why do pieces of artificial satellites return to Earth from time to time? Surely Equation 2.5 suggests that as the satellite slows down, its height h above the Earth will increase. We shall return to this when we have considered work and energy.

Inertia, momentum, work and energy

You cannot do work without energy. The dimensions of work and energy are ML^2T^{-2}, which are those of a force, or a weight, times a distance moved in the direction of the force or weight.

Before the age of nuclear energy (in spite of Einstein's relativity) students were taught the separate principles of the *conservation of energy* and the *conservation of mass*. In the ordinary world it may still be useful to think of them as separate principles, rather than as interchangeable.

A body remains stationary, or moves with a constant speed and direction, unless acted on by a force. This is Newton's First Law of Motion, a restatement of Galileo's. If you wish to set a stationary object in motion, or to slow or stop a moving object, you must apply a force. When you do this, Newton's Third Law says that to every force there is an equal and opposite reaction. The reaction to the force that starts, stops or slows a body is the body's *inertia.* Inertia is a property of the body and only appears as a force in reaction to an applied force. The body's mass is the property that determines inertia, and this is independent of position. It will be the same on the Moon as on the Earth.

Momentum is the product of a mass (m) and its velocity (V) [MLT^{-1}], and is therefore a vector. Momentum is conserved, so if an asteroid strikes the Earth (and remains embedded), the new momentum is the sum of the momenta just before impact. The Earth's mass is very nearly 6×10^{24} kg, and its speed in its orbit is about 30 km s^{-1}, so its momentum is nearly 180×10^{27} kg m s^{-1}. A few sums around these figures will show you that a collision of this sort would have to be with something rather large to have any great effect on our orbit – but it would have a large local effect. Hubbert (1937: 1505) looked at the requirements for modelling such an impact.

Acceleration (a) is the rate of change of velocity with time, so force is the time rate of change of momentum:

$$d(mv)/dt = dv/dt = ma = F, \tag{2.6}$$

assuming the mass to be constant (i.e. this will not be the right equation for the momentum of very small particles travelling nearly at the speed of light). As mentioned earlier, acceleration is not necessarily in the direction of motion. A satellite, as we have seen, has an acceleration normal to its orbit, towards the centre of its planet.

If the motion of a body is about an axis, as a wheel turning, the Earth rotating, or a satellite in orbit, the body will have an angular velocity (in radians/second), perhaps angular acceleration (rad s^{-2}), angular inertia and angular momentum. Angular inertia is the property of a rotating mass to react to a turning force or *torque*. Torque is a product of the force and the distance of its application from the axis of revolution, and has the dimensions $(MLT^{-2})(L) = ML^2T^{-2}$, which are the dimensions of *work*, or in units, newton metres (Nm). A large force acting on a short lever does the same work as a small force on a long lever (Fig. 2.3). This is how you lift a car with a jack, and how some weighing scales work. You can tighten a nut more easily with a longer spanner. The *moment of inertia* is a measure of resistance to angular acceleration and again, it is a passive property of a revolving system $[ML^2]$. Angular momentum, or the moment of momentum, is the product of the momentum and the length of the radius of rotation $[ML^2T^{-1}]$, in units of kg m^2 s^{-1}.

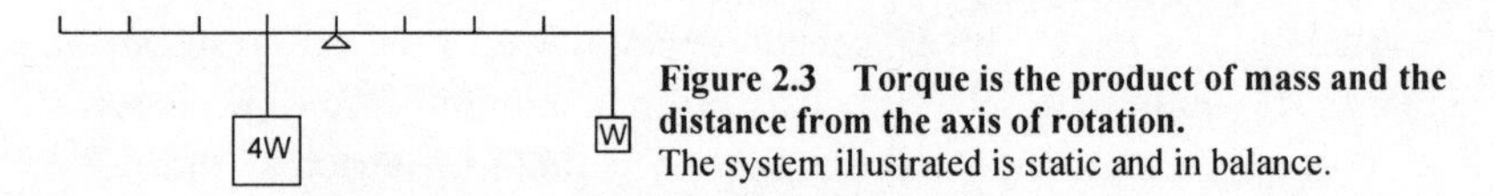

Figure 2.3 Torque is the product of mass and the distance from the axis of rotation.
The system illustrated is static and in balance.

Why does a spinning skater spin faster when her arms are brought closer to the body? The conservation of momentum, in this case angular momentum, means that if you reduce the moment (the radius of rotation) you increase the speed of rotation. The same is true of the Earth–Moon system.

All these quantities are gathered together in Table 2.1.

Table 2.1 Summary of quantities involved in concepts of energy and motion.

Quantity	Symbol	Units	Dimensions
energy	E	J = N m	ML^2T^{-2}
work	w	J = N m	ML^2T^{-2}
power = energy/time	P	W = J/s = kg m^2 s^{-3}	ML^2T^{-3}
inertia		kg	M
momentum = mass × velocity	p	kg m s^{-1}	MLT^{-1}
angular velocity	ω	rad s^{-1}	T^{-1}
angular acceleration	α	rad s^{-2}	T^{-2}
moment of force	M	N m = J	ML^2T^{-2}
moment of velocity		m^2 s^{-1}	L^2T^{-1}
moment of inertia	I	kg m^2	ML^2
m. of momentum, angular momentum	$b, p_{\ominus}$	kg m^2 s^{-1}	ML^2T^{-1}
torque	T	N m = J	ML^2T^{-2}

Potential energy

Gravitational potential energy is energy due to position – energy that can be realized by letting the body fall. Fill a hand-basin with water, and the water has potential energy that can be realized by taking out the plug. The water can then do work as it descends to a lower level. Surplus electric energy is used to pump water from the Wivenhoe dam in Queensland, Australia, to the higher Splityard Creek dam. Its potential energy in the Splityard Creek dam means that electricity can be generated at peak periods by letting the water flow back into the Wivenhoe dam through the hydroelectric plant. Potential energy is relative to some datum.

A body of mass m at height h above the ground has potential energy *relative* to the ground of mgh N m because that is the amount of work it could do in falling a vertical distance h to the ground in a frictionless process. In reality, some of the energy is invariably lost by friction to heat. The full supply level of Splityard Creek dam is about 100 m above that of Wivenhoe dam, so the potential energy of the water behind the Splityard Creek dam relative to the water behind the Wivenhoe dam is nearly 10^6 N m, or 1 MN m or 1 MJ or 1 MW s per cubic metre.

Kinetic energy

Newton's First and Third laws of motion imply that any body in motion has energy because of that motion; and that work is required to stop its motion. The dimensions of kinetic energy are the same as for other energy (ML^2T^{-2}) but it is more usefully regarded as a mass times the square of a speed or velocity (rather than as a weight times a distance, as for potential energy). For a body of mass moving at velocity V the kinetic energy is $mV^2/2$. In a frictionless process, this amount of work could be performed.

Earlier we asked why pieces of artificial satellites return to Earth from time to time when Equation 2.5 suggests that as the satellite slows down, its height h above the Earth will increase. It is a matter of energy. The satellite has energy due to its motion in orbit and to its position in the gravitational field of the Earth. Its kinetic energy is

$$E_k = \tfrac{1}{2}mr^2\omega^2 = \tfrac{1}{2}m\,G\,M_E/(R_E+h) \qquad [ML^2T^{-2}]$$

where ω is the angular velocity (in rad s^{-1}: the right-hand side uses Eq. 2.5). The potential energy is found by considering the work required to move the mass in the gravitational field from a distance x from the earth's centre of gravity to $x+\delta x$. This, from Equation 2.2, is

$$\frac{mGM_E}{x^2}\,\delta x\,.$$

The work done in moving the mass from x to ∞, where the potential energy is zero, is

$$\int_x^\infty \frac{mGM_E}{x^2}\,dx = \left[\frac{mGM_E}{x}\right]_x^\infty = \frac{mGM_E}{x}.$$

Hence the potential energy at (R_E+h) is

$$E_p = -\,m\,G\,M_E/(R_E+h)$$

and the total energy is the sum of the kinetic and potential energies,

$$E = E_k + E_p = -\,m\,G\,M_E/(R_E+h).$$

This is negative, so as the kinetic energy decreases and the total energy becomes a larger negative number, so h also decreases. And as h decreases, its speed increases until it enters the atmosphere, where frictional resistance slows it down and it returns to Earth. What meaning is to be attached to negative potential energy? If a satellite is considered to have zero energy when effectively isolated from the gravitational field of its planet, and you have to do work on it to get it there from its orbit, then its potential energy in its orbit is negative (E_p + Work = 0).

Body and surface forces

Forces that act on all the components of a body of matter, such as gravity and magnetism, are called body forces: forces that act on the surface of a body, such as a push against a box and a hammer blow on a nail, are called surface forces or contact forces.

The body force of gravity acts on everything on Earth – indeed, everything that has mass in the universe. Surface forces are rather special. Take sliding, for instance. Rocks may be *pushed* along a surface by a surface force, or they may slide when the shear component of weight exceeds the resistance from friction. Sedimentary rocks compact under the body force of gravity, but some of the grains may deform under surface forces applied by their neighbours.

Buoyancy

Buoyancy is a special problem, and one of importance in geology. Is it a body force or a surface force? It is clearly a surface force when a body like a boat *floats*. It is clearly a surface force if a boat like a submarine sinks – or is it? If the buoyant object is an immiscible fluid, it seems it must be a surface force; but if it is the same fluid at a different temperature, is it a body or a surface force?

When we weigh ourselves on a scale, we are measuring the sum of the masses of all the particles of our body in the gravitational field. However, our atmosphere is also in the gravitational field, and it has a buoyant effect on us, as Archimedes might have guessed when he got out of his bath, and it reduces our weight by the weight of the air displaced by our body. When you weigh something by balancing it with weights, the error due to buoyancy in air is only zero when the density of the object weighed is the same as that of the weights – that is, their volumes are equal. Weights determined on spring scales are always in error by a buoyancy amount that is small for dense objects, but could reach about 0·5 per cent for objects of small density.

In a swimming pool, as we all know, our weight is further reduced because it is reduced by the weight of the water displaced by our body. A body *floats* only when it displaces less than its volume of water; that is, its *bulk* density must be less than the density of water. This is why steel ships float. This principle is used for separating "heavy" minerals from "lighter" (that is the jargon; but we would be more accurate if we used "denser" and "less dense" or "more massive" and "less massive").

Consider the following[2] in order to clarify your thoughts on buoyancy. There is a lake with no inflow and no outflow. A depth gauge is in the water. On the bank there is a dinghy, a rock and a piece of wood. You put the dinghy into the water: what is the effect on the water level of the lake? You get into the dinghy: what is the effect? Someone hands you a stone: what is the effect? You are then handed the piece of wood: what is the effect?

You row out into the middle of the pond and throw the stone overboard: what is the effect? Then you throw the piece of wood overboard: what is the effect? (You are at university now, so only look at the answer given in the Appendix when you are satisfied that you have the correct answer.)

Buoyancy acts also on fluids. When you take a hot shower in winter, the shower curtain tends to be sucked in on you, because the warm water warms the air, which rises because it is now less dense, and the cold tries to flow in to replace it. If a river brings sediment-laden water to a lake, as the Rhône does to Lake Geneva, the dirty, denser, water may flow down the slope to the bottom of the lake, displacing the cleaner water. This is the position of minimum potential energy. The speed at which it flows depends on the density contrast, the slope of the bottom and perhaps its smoothness, and the viscosity of the water – at least. Figure 2.4 shows the essentials, but is rather naïve. If you would like greater depth, see Turner (1973: 178ff.).

Buoyancy is the main driving force of convection because fluids expand when

2. This was probably an examination question, but I have been unable to trace it. I therefore apologize to whoever owns the copyright – but it is too good to omit.

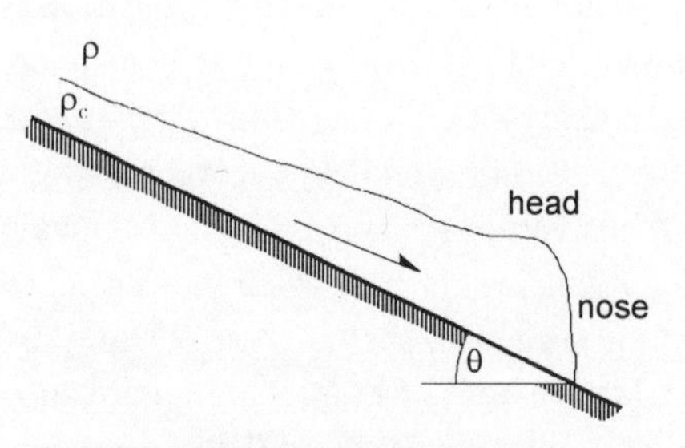

Figure 2.4 Density current.
The head of flow is attributable to the work required to displace the static water.

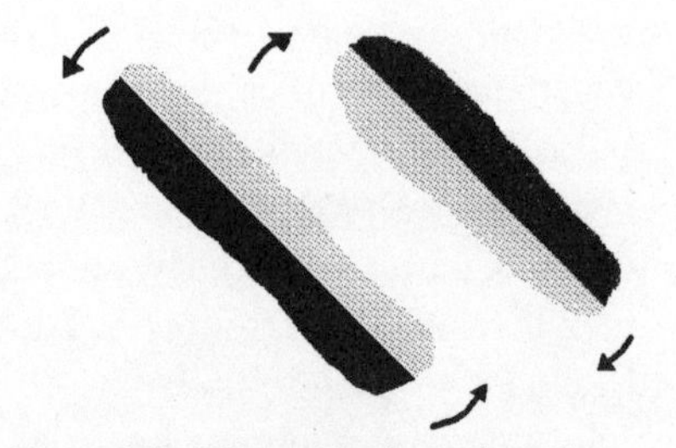

Figure 2.5 Diagrammatic section of a stable and an unstable sequence.
A less dense substance (lighter ornament) overlying a denser substance is stable, and there is a force tending to make them horizontal. Denser over less dense is unstable, and the less dense will tend to rise above the denser.

they are heated, and so become less dense – but the energy may come from an outside source such as the flame under the saucepan. Remembering that in geology we regard as fluids substances that are certainly solids on a short timescale, the force of buoyancy acts on any volume of less dense rock that is overlain by denser rock (Fig. 2.5).

The drill pipe fallacy or paradox In the 1950s, development drilling in several oil fields around the world was hampered by the drill pipe's sticking against formations that had been depleted by production. This was called "wall sticking" or "differential sticking". It was due to the lateral pressure gradient between the mud in the hole and the depleted fluid pressures of the reservoir (Fig. 2.6); and it could hold the pipe against the wall of the hole, cushioned in the filter cake that forms around the hole in permeable rocks, with great force. The stuck point, in oil field jargon, was not usually the bottom of the pipe, at the drill collars, but higher. To free the pipe, one would first pull on it (in those days, not now) – the strength of pull being limited by the tensile strength of the pipe. The question was, how hard could the pipe be pulled without it parting? The answer is not as simple as it seems.

The weight of steel pipe is reduced by the weight of the mud displaced, but

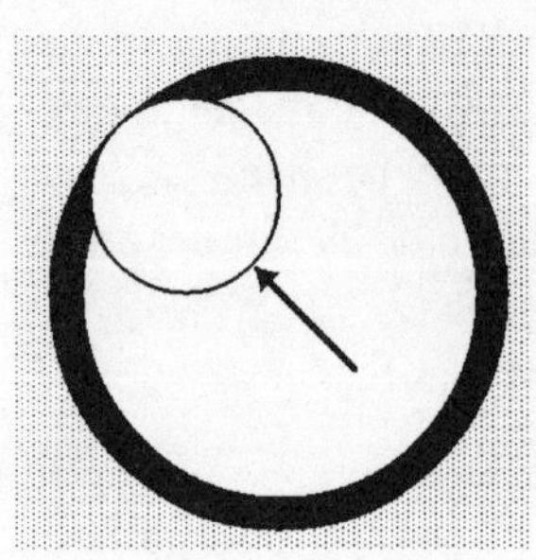

Figure 2.6 Wall-sticking in horizontal section through the borehole.
The pressure of the mud holds the pipe to a depleted reservoir, sealed by the mud cake (cake of mud formed by filtration of the solids in the mud as a result of the greater energy of the mud in the hole).

what is the weight of the pipe above the stuck point? If the force of buoyancy acts on the pipe above the stuck point, the pipe could be pulled that much harder because it is the tension in the pipe that is limited. If the force of buoyancy does not act on the pipe above the stuck point, as was commonly believed at that time, the pipe could only be pulled to its nominal tensile strength (less a safety margin). This problem has been called the "drill-pipe fallacy".

Consider the forces acting on an open-ended pipe hanging freely in a borehole filled with mud. Let us take the following values of the parameters for the pipe and mud: length of pipe 3000m; weight of pipe $263{\cdot}6\,\mathrm{N\,m^{-1}}$, displacing $3{\cdot}33 \times 10^{-3}\,\mathrm{m^3\,m^{-1}}$, in mud of mass density $1500\,\mathrm{kg\,m^{-3}}$; cross-sectional area of the pipe $3{\cdot}3 \times 10^{-3}\,\mathrm{m^2}$.

The pressure exerted by the mud at the bottom of the pipe is $\rho gd = 1500 \times 9{\cdot}8 \times 3000 = 44{\cdot}1 \times 10^6$ Pa, or 44·1 MPa. This is providing an upward force of 145·53 kN on the bottom of the pipe, equivalent to the weight of about 550 m of the pipe in air. Drill pipe is stacked in the mast or derrick of a drilling rig in lengths of about 30 m. These pipes have a visible bend, so it is clear that 550 m could not stand on end but would buckle. But it does not buckle in the borehole, so either the physical interpretation is wrong or there are pertinent factors that have not been taken into account.

Alternatively, the 3 km of pipe displaces $9{\cdot}9\,\mathrm{m^3}$ of mud that weighs $1500 \times 9{\cdot}9 \times 9{\cdot}8 = 145\,530$ N or 145·53 kN, so the weight of the pipe in the mud is $790{\cdot}86 - 145{\cdot}53 = 645{\cdot}33$ kN. The *effective* weight is $215{\cdot}1\,\mathrm{N\,m^{-1}}$, and the tensile load on the pipe varies continuously from its total effective weight at the top to zero at the bottom, the whole length of the pipe being in tension. Buoyancy, like weight, must be a body force – the difference between the effects of the body force of gravity on materials of two different densities. (Many will disagree with the resolution of this paradox, attributing the result to compressibility of the material under surface forces, but it makes you think.)

Time and force

Force, as we have seen, is a mass times its acceleration and has the dimensions MLT^{-2}. The question is, what is the effect of the dimension of *time* on the effects of force? The *coefficient of viscosity*, which is a measure of the internal friction of fluids, resisting flow, has the dimensions of $ML^{-1}T^{-1}$, which may be thought of as a force or stress multiplied by a time. What is the effect of the dimension of time? An interesting part of the answer to these questions lies in the theory of modelling. If we wish to model a process that takes millions of years so that its effects can be seen in days, we must use materials of smaller viscosity in the model. In other words, small forces applied over very long periods can have the same effect as larger forces applied over shorter periods. We have natural examples in the pitch lakes, as mentioned in the discussion of solids and fluids above (p. 7).

Coriolis force

Why does the carousel attendant at the funfair have to lean over in order to walk radially outwards from the centre? Why do the ocean currents in the same hemisphere circulate the same way, but those in opposite hemispheres the opposite way? Why does a cyclone in the Southern Hemisphere rotate clockwise when viewed from above, but anti-clockwise in the Northern Hemisphere?

Consider two dimensions first. If you want to walk in a straight line radially on a revolving stage (that is, the path drawn on the stage would be a straight line), one of the forces acting on you gives you an acceleration as the distance from the centre increases. Your angular velocity remains constant, but your speed increases as the radius increases. So if the stage were rotating to the right as you look out from the centre, the acceleration in that direction means that you would have to lean over to the *right* to counteract it. Failing that, your path will deviate to the left of the straight line.

What about bowling a ball along the stage from the centre? Assuming a frictionless process, the ball would indeed travel in a straight line out from the centre, but its path over the stage would be curved.

If you had a very-long-distance gun and wished to fire a shell from Brisbane to Canberra (which is nearly due south of Brisbane), you would miss if you pointed the gun directly at Canberra because Brisbane is travelling a greater distance around the world's axis in a day than Canberra does. Both have an angular velocity of 15° an hour, but this is a speed of 1480 $\mathrm{km\,h^{-1}}$ in the latitude of Brisbane (27·5°S) and 1366 $\mathrm{km\,h^{-1}}$ in the latitude of Canberra (35°S). The Brisbane component of the shell's velocity would carry it to the east of Canberra – to the left. If Canberra returned the fire, its component of the shell's velocity due to latitude would be less than the velocity of Brisbane, and so the shell would fail to keep up with Brisbane and land to the west of Brisbane – again, missing to the left. So for any longitudinal motion in the Southern Hemisphere there is a tendency for the path to deviate to the left. This is why the winds blowing out of a high-pressure area are deflected to the left, giving an anti-clockwise rotation. Winds blowing into the centre of a cyclone in the Southern Hemisphere also deviate to the left and give rise to a clockwise motion. In the Northern Hemisphere, the deviation is to the right.

Equilibrium and stability

A body at rest is in mechanical equilibrium, but there are three sorts of mechanical equilibrium: neutral, stable and unstable. If a body in any position has balanced forces acting on it whatever the position, it is said to be in *neutral* equilibrium (e.g. a cube on a table). *Stable*, literally, means "able to stand". Stability does not mean quite the same thing as equilibrium. If a body oscillates (or would oscillate) for a finite time until it ceases to move, it will eventually reach *stable* equilibrium (e.g. a pendulum). If a body remains static only within very close limits of its position, it is in *unstable* equilibrium (e.g. a pencil standing on its end). A mechanical system is stable if, left to itself, there will be no appreciable change in its state with time. Artificial satellites are in stable orbit if they will remain in their orbit for the length of time required for their purpose.

In geology, with time such an important factor, we are concerned with stability on a slightly more subtle scale. For instance, a horizontal, layered, sequence in which the densities of the rocks increase with increasing depth is mechanically stable because the arrangement minimizes the potential energy of the system. There are no gravitational forces tending to disturb it. If, however, there is a layer of smaller density interposed, the sequence above it (and below it to some extent) becomes unstable because the potential energy of the denser layers above is greater than the potential energy of the less dense layer and there will be a force tending to displace the less dense layer to a position above the denser layers, so minimizing the potential energy of the sequence. This is the physics behind diapirism. A layer of salt, if less dense than the overlying layers, will tend to flow first into ridges, then into "domes", "plugs" or "stocks" that may acquire an almost circular horizontal section. With time they may reach the surface, as has happened in Iran, or very close to the surface, as in Louisiana, USA.

Gravity

Fields

Fields are difficult to define. Within a body of water there is a pressure field. If the water is of constant density and is at rest, surfaces of equal pressure ($\rho g z$) are horizontal (that is, normal to the vector $-\boldsymbol{g}$). When the water flows, the pressure field changes because the *energy* of the water changes. Energy is also a field. Fields may consist of scalars (magnitude), such as temperature, pressure, or energy, or of vectors (magnitude and direction), such as force; and they can be mapped.

The space in which terrestrial magnetism acts, and the space in which gravity acts, are called *magnetic fields* and *gravity fields* respectively. An electric current in a wire sets up a magnetic field around the wire. Fields in physics are spaces (or areas) in which some agent (commonly a force) operates. There are fields at various scales: regional and local. The local magnetic field is of interest to geologists because ore bodies may give rise to a magnetic anomaly. Take gravity.

On the surface of the Earth the force due to gravity acting on any object is, as we saw on page 15,

$$F = -\mathbf{g}\, m, \tag{2.7}$$

g being a vector, the negative sign indicating that it acts downwards. The gravity field is $-\mathbf{g}$ and the force acting on any mass in the field is simply that mass multiplied by the value of the field in that position.

Regionally, it may be sufficient to consider the gravity field as being everywhere $-9{\cdot}8\ \mathrm{m\,s^{-1}}$ directed towards the centre of the Earth, approximately, but in detail, $-\mathbf{g}$ is constant neither in quantity nor direction. We have noted above that there is a latitude effect. A formula for the sea-level value for **g** in latitude ϕ is

$$\mathbf{g} = 9{\cdot}780\,491 + \frac{\sin^2\phi}{189{\cdot}0931} - \frac{\sin^2 2\phi}{169\,491}, \tag{2.7a}$$

which is not considered to be quite correct but is nevertheless the standard. The discrepancy is very small, and of no significance over the limited area of most surveys. The acceleration on the Equator due to centrifugal force, V^2/R, is about $34\ \mathrm{mm\,s^{-2}}$.

The rotation of the Earth about the Sun also affects the value of **g**, as does the Moon's rotation about the Earth, both of which we see in tides (which will be examined more closely below). On a local scale, the topography and elevation above sea level must also be taken into account because hills distort the local gravitational field, as do valleys, and elevation implies extra mass beneath a gravimeter – the instrument for measuring the value of **g**. When all these effects have been taken into account, there are still local variations that are significant to Earth scientists owing to the variations of mass density in the rocks below the surface. The field has to be mapped by careful measurement, and anomalies in the field may indicate geological features of interest.

Universal gravity

When discussing weight, mass, and density (p. 15) we stated Newton's Law of Gravitation: that every particle of mass m_1 attracts every particle of mass m_2 with a force that is directly proportional to the product of their masses and inversely proportional to the square of the distance between their centres of mass, and this force acts along a straight line joining them:

$$F = G\,(m_1\, m_2)/r^2, \tag{2.8}$$

where r is the distance between their centres of mass, and G is the universal constant of gravitation. And we noted that G is not a true constant because it has dimensions:

$$[G] = [MLT^{-2}]\,[L^2]\,[M^{-2}] = M^{-1}L^3T^{-2},$$

which are those of the force of attraction times the square of the distance between the bodies, divided by the product of their masses (N m^2 kg^{-2}).

Sea level

We have discussed terrestrial gravity fields above in sufficient detail for our purposes, but there is another aspect of importance to geologists. It might be supposed that the surface of the seas of the world would present us with a reasonably regular figure that would reflect terrestrial gravity accurately. Presumably it does, but not in a way that makes this obvious.

Mean sea level A fluid is a substance that yields at once (on a short timescale) to the slightest shear or tangential stress. This is why the free upper surface of a body of water at rest is horizontal. We might therefore assume that the mean sea surface around the world describes a regular figure such as a spheroid (a form of ellipsoid). It doesn't quite. There are departures up to about 100 m from the best-fitting ellipsoid.

In any one place on the coast, you can *estimate* mean sea level from tidal records. These records include influences other than those of the Sun and Moon because wind and atmospheric pressure can change sea levels. By damping out short wavelengths, higher frequencies, the change of sea level with time can be recorded. Over a sufficiently long period of time, mean sea level can be approached. How do we relate that mean sea level in one place with the mean sea level in another place?

Figure 2.7 shows the contours on the mean sea surface of the Indian and Pacific Oceans obtained from satellite radar altimetry (the measurement of height by satellite-borne radar, interpreted as the shape of the ocean surface below the known orbit). It is clear that the deviations from the mean ellipsoid are significant, but it is not clear why they are where they are. There is no evident relationship with plates or continents. If the high levels of New Guinea (for example) were to drift west (as the magnetic field does) towards the low levels south of India they would cause significant changes. The total range is from –104 to +79 m.

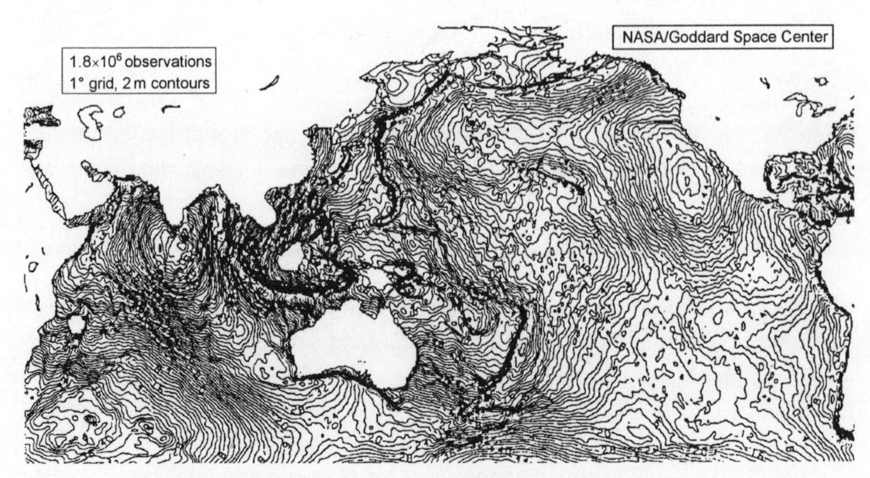

Figure 2.7 Mean sea-surface topography of the Indian and Pacific oceans.
South of India, the geoid surface is 104 m below the best-fitting ellipsoid; in New Guinea, 79 m above. (Courtesy NASA/Goddard Space Center).

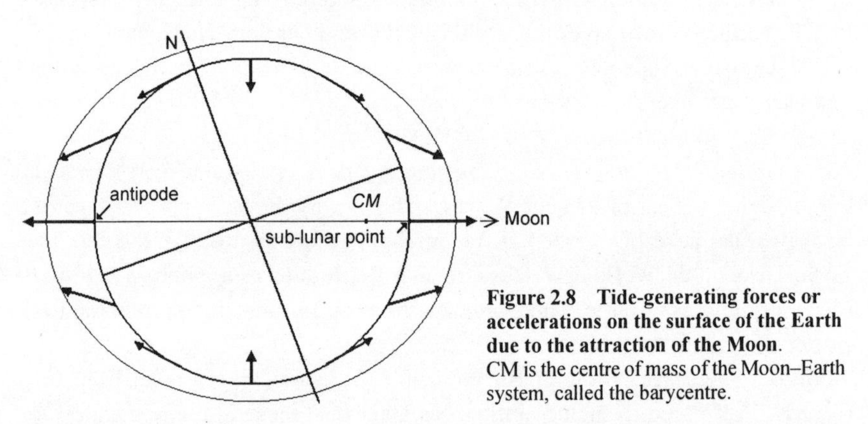

Figure 2.8 Tide-generating forces or accelerations on the surface of the Earth due to the attraction of the Moon.
CM is the centre of mass of the Moon–Earth system, called the barycentre.

Tides The waters of the Earth are subject to various forces: centrifugal forces from the Earth's rotation in orbit, rotation about its axis, and rotation with the Moon; and gravitational attraction to the Earth, the Moon and the Sun. When the Moon is new, and it is pulling in the same direction as the Sun, tides are higher than when there is a half-Moon. That is fairly straightforward; but why are there very high tides soon after the Moon is *full*, on the opposite side of the Earth from the Sun? Indeed, why are there *two* tides a day in most places, and not just one?

The centre of mass of the Earth–Moon system, called the barycentre, is not at the centre of the Earth, but displaced towards the Moon (still within the Earth, at about 0·74 R_E from the centre, R_E being 6371 km) (Fig. 2.8). The Moon and the Earth rotate around their barycentre, and at the centre of the Earth the centrifugal force due to rotation about the barycentre exactly equals the gravitational attraction between the Earth and the Moon – indeed, there is a circle around the Earth on which the Moon is on the horizon and in which these two forces balance each other. Away from that circle, the balance is not found, and this imbalance gives rise to *tide-generating forces* or *tide-generating accelerations*. The centrifugal force exceeds the gravitational on the side away from the Moon, and the gravitational exceeds the centrifugal on the side towards the Moon. More importantly, the water on the side of the Earth away from the Moon is attracted less by the Moon than the water nearest it. The oceans are drawn out (or an ideal ocean would be drawn out) towards and away from the Moon, and the Earth rotates inside giving roughly two tides each day in most places. The point on the Earth's surface that is vertically beneath the Moon is called the sublunar point; its counterpart on the side away from the Moon is called the antipode.

Consider the attraction of the Moon only. The distance between the centre of the Earth and the centre of the Moon, d_M, varies from about 363 000 km to 406 000 km, averaging about 384 400 km. This is a little more than 60 R_E. So, as an approximation, the attraction is proportional to $1/60^2$ at the centre of the Earth, $1/59^2$ at the sublunar point and $1/61^2$ at the antipode.

The acceleration of unit mass at the Earth's centre due to the Moon is

$$a_c = G\,M_M/d_M^{\,2}\,;$$

at the sublunar point it is

$$a_s = G\,M_M/(d_M - R_E)^2\,;$$

and at the antipode it is

$$a_a = G\,M_M/(d_M + R_E)^2\,.$$

So the differences are, after some simplifying (see Note 2 for help on this if you need it),

$$(a_s - a_c) \approx 2G\,M_M\,R_E/\,d_M^3.$$

and

$$(a_a - a_c) \approx -2G\,M_M\,R_E/\,d_M^3.$$

The force vectors in the sublunar and antipodal points are therefore towards and away from the Moon, and they are normal to the sea surface at those places.

On the surface of the Earth between the sublunar and the antipodal points there are lunar tide-generating forces, and the force vector towards or away from the Moon has a transverse component, called the *tractive* force (Fig. 2.9). There is a great circle around the Earth between the sublunar point and the antipode in which the tractive forces are zero and the force vector is vertically towards the centre of the Earth. The tractive forces tend to move the water towards the sublunar point and the antipode.

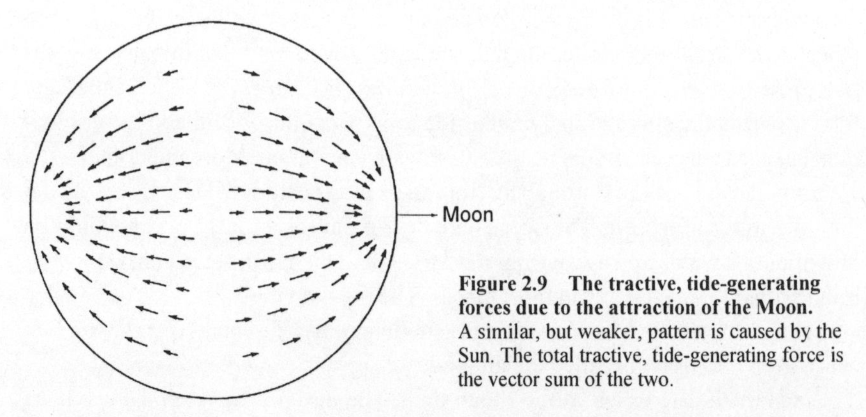

Figure 2.9 The tractive, tide-generating forces due to the attraction of the Moon. A similar, but weaker, pattern is caused by the Sun. The total tractive, tide-generating force is the vector sum of the two.

The Sun gives rise to similar but smaller tide-generating forces, and the net effect is the vector sum of the two. The effect of the Sun is a little less than half the effect of the Moon, but it is not negligible. So, at full Moon and when there is no Moon, the Sun and Moon effects are added, and we get higher high tides and lower low tides than at other times. These are called *spring tides*. When the Moon is half and the Sun and Moon are at right angles to each other, the high Moon tides are reduced by the low Sun tides, and vice versa, and the high tides are lower and the low tides are higher than at other times. These are called *neap tides*.

The Earth rotates within the tidal envelope so created, and friction (in shallow water particularly) and the obstruction of the continents distort the rhythm of the tides significantly – and tend to slow the rate of the Earth's rotation, lengthening the day slightly, *very* slightly. The frictional resistance causes a slight delay in the tides, so that high tide typically comes after the Moon's passage across the meridian. Oceanic tidal ranges are about a metre: in shallow waters marginal to large oceans, they may be 10 m or more. Within seas that are virtually isolated from the ocean water movements, such as the Mediterranean and the Black Sea, the tides are very small.

It will be clear from Figure 2.8 that two consecutive high tides, or two consecutive low tides, will only be equal in height when the Moon is over the Equator. At other times, the Earth's rotation about its axis will bring most parts of the Earth to areas of different tide-generating forces.

There are also tides in the atmosphere and in the solid Earth, but the latter are very small and very difficult to distinguish from the slight deformation caused by the oceanic tides.

Mountains

As early as 1735, a Frenchman by the name of Pierre Bouguer (1698–1758) found that the Andes mountains in South America did not attract the plumb bob or pendulum as much as he expected. (How did he know what to expect?)

In the 19th century, topographic survey of the northern part of the Indian subcontinent revealed discrepancies between the positions obtained by triangulation and those by astronomical observation corresponding to a deflection of 5·24″ (*seconds*) of arc. In 1855, Archdeacon Pratt studied the discrepancies and came to the conclusion that they were about one third the size of those expected (15·88″). And in France at about the same time, it was found that a plumb bob was deflected *away* from the Pyrenees, *towards* the Bay of Biscay.

What is to be made of these observations? In the first place, they were expecting the plumb bob to be attracted from the true vertical by the mass of the mountains to one side. From the shape, volume and density of the nearby mountains the mass and centre of mass can be estimated; and from this the force can be estimated. This force does work in *lifting* the plumb bob from its true vertical position. If the attraction is less than expected, then the mass of the mountains is less than expected. If it can then be shown that the error of estimate of the mass above the level of the plumb bob is smaller than that implied by the discrepancy, then the deficiency of mass must be *beneath* the mountains, *below* the level of the plumb bob. This was the conclusion of both Pratt and Airy in 1855, but Airy explained it in terms of the mountains floating on a more dense layer. This was the beginning of the principle of *isostasy*. Isostasy is comparable to the floating of an iceberg. The word strictly implies mechanical equilibrium, but equilibrium is most unlikely in mountainous regions by their very nature, particularly the erosion and removal and displacement of mass from them.

3
OPTICS

Light is electromagnetic radiation. In some respects it behaves as waves; in others, particles. Its speed in space, c_o, is 299 792·4 km s^{-1}, as is the speed of other types of electromagnetic radiation. This is an absolute constant, but light travels rather more slowly in air, and much more slowly in transparent materials such as crystals, glass and water. Radio waves are near one end of the range of electromagnetic radiations (wavelength λ greater than a few centimetres): γ-rays are near the other end (λ less than 100×10^{-15} m). Within the visible range of light, there is a *visible spectrum* ranging from the longer-wavelength red to the shorter-wavelength violet, through orange, yellow, green, blue, and indigo. λ_{red} is about 700×10^{-9} m or 700 nm; λ_{violet} is about 400 nm. Outside this visible spectrum are infrared and ultraviolet. Infrared can be detected on photographic film, and by its heat: ultraviolet can be detected because it excites light in the visible spectrum from some minerals and organic substances.

The same relationship between wavelength, λ, speed, c, and frequency, ν (the Greek letter nu), exists in electromagnetic radiation as in other waves, that is

$$\nu = c/\lambda, \qquad (3.1)$$

where c is the speed of light (but not necessarily as large as c_o). However, electromagnetic radiation does not always behave as waves. In that case, there is a relationship between the photon quantum energy, E, the frequency of light, and the Planck constant (h),

$$E = nh\nu \qquad (3.2)$$

where n is an integer. This is the basis of the quantum theory initiated by Max Planck in 1900 – but we shall not go into it.

Black objects are black because black absorbs practically all wavelengths that fall on it. White objects are white because white reflects practically all wavelengths that fall on it.

Reflection and refraction

When we look in a mirror, we see a reflection of ourselves. We can go further and measure the angles of reflection by setting a mirror vertically on a piece of paper on a table and measuring the positions and apparent positions of pins (Fig. 3.1). This is called geometrical optics. Energy radiates from a source – in this case, the pin – and the lines along which the energy radiates in different directions are called rays. The rays are perpendicular to the wave front of the radiating energy. When they strike a surface, the atoms at that surface also emit radiation of the same frequency. The basic law of reflection, which can easily be established by the experiment shown in Figure 3.1, is that the *angle of incidence is equal to the angle of reflection.* The angle of incidence is in the plane of incidence, which is normal to the reflecting surface.

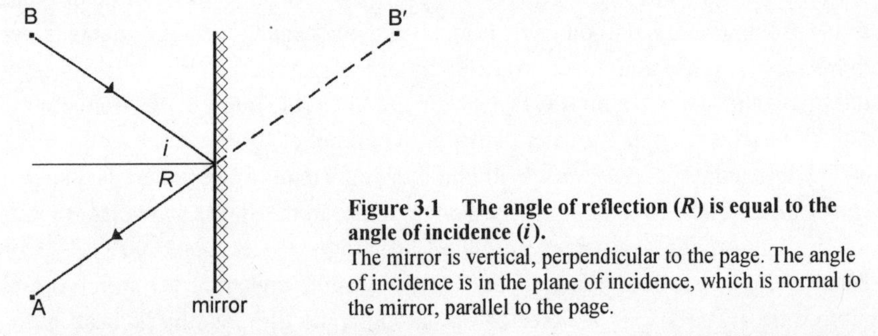

Figure 3.1 The angle of reflection (R) is equal to the angle of incidence (i).
The mirror is vertical, perpendicular to the page. The angle of incidence is in the plane of incidence, which is normal to the mirror, parallel to the page.

When we look at people standing in a swimming pool, their submerged parts are distorted. If we dip the end of a straight stick in water at an angle, it appears to bend upwards at the surface of the water. This is *refraction.* Looking along the stick, the apparent angle increases from zero when the stick is held vertically, becoming very large as we rotate the stick from the vertical before we have to abandon the experiment (Fig. 3.2a). Thus both the reflected and the refracted rays change direction at an interface, but the frequency remains the same. Viewing the sun rising and setting from near sea level, its upper limb touches the visible horizon when it is actually about 34′ below the horizon because the rays have been bent by refraction on their passage through the Earth's atmosphere. The speed of light is greater in the rarefied air at higher altitudes. Mirages are similarly caused by refraction, the hot air close to the ground being less dense that the cooler air higher up.

By convention, we measure the angles from the normal to the surface. Snell's law tells us that

$$\frac{\sin i}{c_{\mathrm{i}}} = \frac{\sin R}{c_{\mathrm{i}}} = \frac{\sin r}{c_{\mathrm{r}}} \tag{3.3}$$

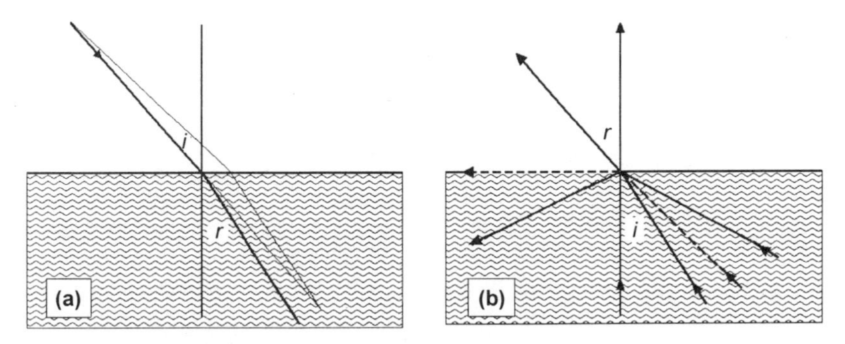

Figure 3.2 Refraction.

(a) The angle i is the angle of incidence, the angle r, the angle of refraction, and the incident medium is less dense than the refracting medium. If you look along a partly submerged stick inclined at an angle i to the vertical, the submerged part appears to be above its real position.

(b) When the incident medium is denser than the refracting medium, the angle of refraction is larger than the angle of incidence, and there is a critical angle at which the refracted ray is in the plane of separation of the two media (dashed line). Angles of incidence greater than this result in total reflection.

Both (a) and (b) are incomplete because incident light is both refracted and reflected, with two exceptions discussed in the text, at the interface between different transparent media.

where i is the angle of incidence, R that of reflection and r that of refraction; and c_i is the speed of the incident energy or wave, and c_r is the speed of the refracted wave. In optics, the ratio c_i/c_r is called the refractive index, with the symbol n, when light passes from a vacuum into a denser medium ($c_i > c_r$). We shall pursue this shortly.

These two laws were unified by the French mathematician Pierre de Fermat (1601–65) in the middle of the 17th century by the *principle of least time* – that the light we see was the light that followed the path that would reach our eyes in the shortest time. This implies that light travels at different speeds in different media, slower in water than in air, for example. In Figure 3.1 we see that pin B has its mirror image at B′, and that the shortest distance from B′ to A is a straight line. The shortest distance from A to B that includes reflection from the mirror is the path that starts towards B′ and is reflected to B. Looking at this the other way round, if a point of light were to be placed in position B, an observer at A unable to see B directly would not know whether the light was at B or B′. The *principle of reciprocity* requires that light travelling in the reverse directions traverses the same paths. (Do not forget that light is going out from the point in all directions, but we only see that which arrives first.)

In Figure 3.3, the light travels from a plane source AA′. When the ray from A reaches the interface at B, the remainder of the plane wave front BC still has some distance to travel in air while the ray at B continues in the other medium. The frequency of the plane wave remains the same, but its speed changes, so its wavelength also changes. In the time taken for the incident ray to travel from B′ to C′,

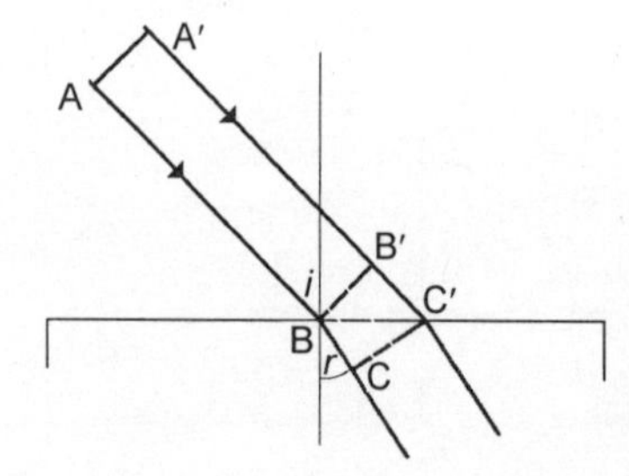

Figure 3.3 A geometrical construction for the refraction path, in section.
When the wave front AA' reaches BB', it begins to be refracted. The velocity in the lower medium is less than that in the upper, so the wave only travels the distance BC in the time it takes the wave at B' to reach C'.

the refracted ray travels a shorter distance (in this case) from B to C, and the direction of the wavefront is altered. The refracted wave is parallel to CC′.

When the incident light is in the medium with larger refractive index, smaller speed $c_i < c_r$ (Fig. 3.2b), the angle of refraction is larger than the angle of incidence. As the angle of incidence increases, so the angle of refraction increases until the refracted ray is in the plane of contact between the two media and the angle of refraction is 90°. This angle of incidence is called the critical angle, i_c, and at any angle of incidence greater than i_c the ray is totally reflected. You can see this when diving. Looking upwards, you can see things above the surface; but there is an angle from the vertical greater than which the surface appears as a mirror – as indeed it is. Snell's law tells us that $i_c = \sin^{-1} c_i/c_r$ where c_i is the speed of light in the denser medium and c_r that in air. This is the inverse of the relationship that gives the refractive index, so $i_c = \sin^{-1} (1/n)$. For water with a refractive index of 1·333, the critical angle is about 48·5°. For very precise work we would have to take into account that the speed of light, c, is not quite the same in air as in a vacuum (the refractive index of air is 1·0002).

Snell's Law applies to the reflection and refraction of sound and seismic waves as well (see p. 100).

The *coefficient of reflection* is the ratio of the intensity, I, of the normal incident to the normal reflected light (the basic SI unit of luminous intensity is the candela, cd):

$$\mu_r = I_R/I_i.$$

In geology, we are also concerned with the *reflectance* of coal-like material, particularly vitrinite, in studies of coals and the diagenesis of sedimentary rocks. Reflectance, R or ρ, is the ratio of reflected radiant flux, which is the time rate of reflected radiant energy ($W = J\,s^{-1}$), to the incident radiant flux:

$$R = \phi_R/\phi_i. \quad (3.4)$$

Refractive index

As mentioned above, the refractive index is the ratio of the incident wave speed to the refracted wave speed for light passing from a vacuum into a denser medium. It is also the ratio of the sine of the angle of incidence and the sine of the angle of refraction,

$$n = c_i / c_r = \sin i / \sin r.$$

However, the amount of refraction, and so the refractive index, also depends upon the wavelength of the light, the refractive index increasing as wavelength decreases. The variation of refractive index with wavelength is called *dispersion*, and *chromatic aberration* in the context of lenses because different colours focus on slightly different planes (many camera lenses have a mark for infrared focus at ∞, which is less than the scale for normal light at ∞). White light, being a mixture of colours and therefore of frequencies and wavelengths is split into a *spectrum* of colours by its passage through a prism. The light is refracted on entry into the prism, and again on exit, with violet refracted most, then indigo, blue, green, yellow, orange, to red. With yellow light with a wavelength of 589 nm, typical refractive indices are: air 1·0002, water 1·333, glass about 1·5, Canada balsam 1·54, diamond 2·417. It is its very high refractive index that gives the cut diamond its sparkle. Calcite and quartz are examples of minerals that have two refractive indices, depending on the orientation of the crystal. This is called birefringence and it will be discussed when we have considered polarization of light.

Interference

Interference is produced when two light sources are superimposed. If two wave trains of the same wavelength and in the same phase, but of different intensity, are superimposed, a new wave train of the same wavelength and phase is formed, but of greater intensity. If the two phases differ, both the amplitude and the phase of the new wave will be different. If the wavelengths differ, the new train will be a mixture of wavelengths – that is, a mixture of colours. We saw this in the context of water waves in Figure 1.1.

Thin plates and films – for example, a soap bubble – give interference colours. Each ray of light, on reaching the surface of the film, is partly reflected and partly refracted. The refracted ray, on reaching the other surface of the film, is partly reflected and partly refracted. The reflected ray returning to the first surface is there partly reflected and partly refracted, and the refracted ray now coincides with a reflected ray at that point. It can be shown that the angle of reflection of the external ray and the angle of refraction of the internal ray are such that these two now coincide. However, the internal ray has travelled farther (and slower) and the two are out of phase. The amount of the phase difference is a function of the thickness of the film. If light of a single wavelength is used, interference results in a series of lit and dark interference fringes (where the light is in phase and out of phase respectively) the spacing of which depends on the film or plate thickness. In the case of white light, the position of the fringes depends on the wavelength, so coloured interference fringes are observed.

Polarization

The electromagnetic waves in monochromatic light oscillate at a definite frequency, but the directions of oscillation in different planes normal to the direction of propagation are not fixed. There is no preferential direction of oscillation. If such light passes through a crystal of tourmaline that has been cut parallel to its principal axis, it is transmitted whatever the orientation of the crystal. It is found, however, that the transmitted light after passing through the crystal differs from the incident light on the other side in having its directions of vibration restricted to one by passage through the crystal. When there is only one direction of vibration of light, the light is said to be *polarized* or *plane polarized* or *linearly polarized.*

When incident light falls on the surface of a transparent medium, part is reflected, part refracted. Light reflected from a surface is also polarized or partly polarized, and this is why *Polaroid* ™ sunglasses reduce the glare of sunlight on a wet road. They also remove much of the reflected light from water allowing a clearer view into the water. (Polaroid™ is a manufactured preparation of crystals of iodosulphate of quinine, herapathite, all orientated in the same direction in a film of celluloid, plastic or glass.) Maximum polarization of reflected light is achieved when the reflected ray is normal to the refracted ray (Brewster's law). It can be shown quite simply if you examine the geometry of Brewster's law that the tangent of this angle of incidence is the refractive index. So for water, this angle is $\tan^{-1} 1{\cdot}333 = 53°$. If the incident light at Brewster's angle is unpolarized only the light oscillating normal to the plane of incidence is reflected; the rest is refracted and partly polarized in the plane of incidence. If the incident light is polarized in the plane of incidence, none is reflected and the refracted light is polarized in the plane of incidence. If it is polarized normal to the plane of incidence, it is partly reflected at all angles of incidence and the refracted light remains polarized normal to the plane of incidence.

Fresnel's reflection formulae for the proportion of light reflected at various angles of incidence are, $-\tan(i-r)/\tan(i+r)$ for light polarized in the plane of incidence, and $-\sin(i-r)/\sin(i+r)$ for light polarized normal to the plane of incidence. These proportions are in terms of amplitudes of the light. When the incident light is polarized in the plane of incidence and the angle of incidence is Brewster's angle ($i+r=90°$), no light is reflected. The coefficient of reflection is the ratio of the *intensity* of the reflected light to the intensity of the incident light, and the intensity is proportional to the square of the amplitude. So the coefficients of reflection for light polarized in the plane of incidence is $\tan^2(i-r)/\tan^2(i+r)$, and for light polarized normal to the plane of incidence, $\sin^2(i-r)/\sin^2(i+r)$. For unpolarized light, there is another formula due to Fresnel:

$$\tfrac{1}{2}\{[\sin^2(i-r)/\sin^2(i+r)]+[\tan^2(i-r)/\tan^2(i+r)]\}.$$

The coefficient of reflection for normal incident light on a transparent surface can be derived from these formulae and is given by

$$\mu_r=(n-1)^2/(n+1)^2$$

where n is the refractive index. The derivation uses the fact that when the angles i and r or $(i-r)$ and $(i+r)$ are very small, $\sin^2(i-r)/\sin^2(i+r)\approx\tan^2(i-r)/\tan^2(i+r)$ $\approx(i-r)^2/(i+r)^2=(n-1)^2/(n+1)^2$.

The blue sky is partly polarized, most strongly polarized at right angles to the direction of the Sun. This can be verified by looking at the blue sky through a Polaroid™ filter (or one lens of Polaroid™ sunglasses) and rotating it. The amount of light passing through the lens will vary in different parts of the sky and with different orientations of the Polaroid™ filter. The Vikings knew this and used the polarizing mineral cordierite for navigation (even on a cloudy day, the position of the Sun can be estimated to ±2·5°, it is said). Bees can also detect polarized light and use it for navigation.

We cannot distinguish the directions of vibration of light without assistance. If we pass ordinary light through a crystal of tourmaline that has been cut parallel to its principal axis, and then let it pass through another similar crystal, we find that the light emerging from the first crystal is polarized parallel to its axis. We use the second crystal to *analyze* the light emerging from the first, which we call the *polarizer*.

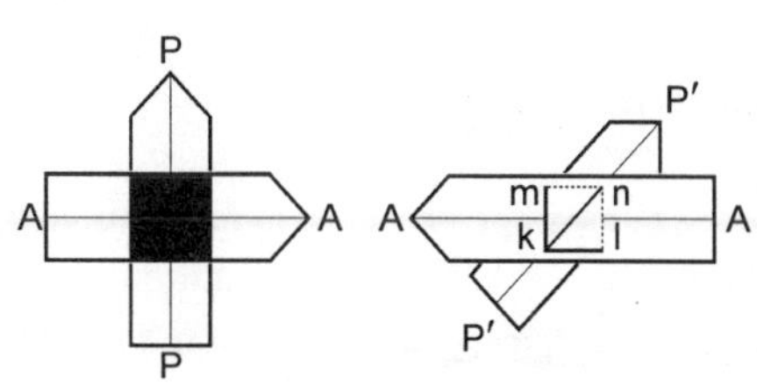

Figure 3.4 Polarizer (P) and analyzer (A), crossed and inclined. Light coming from below (from the other side of the paper, as it were) passes first through the polarizer, which polarizes the light parallel to its principal axis PP. This polarized light cannot pass through another crystal with its principal axis normal to the polarizer (on left in diagram). At other angles, this polarized light can be resolved into two components, *km* and *kl*, on the right figure, of which only *kl* can pass through the second crystal.

In Figure 3.4, light transmitted from below is polarized parallel to PP, so when the analyzer is oriented at right angles to the polarizer, no light is transmitted through the analyzer because it can only transmit light polarized parallel to AA. When the crystals are inclined, the vibrations parallel to P′P′ can be resolved into two directions normal to each other. The amplitude parallel to the axis of the polarizer can be considered proportional to *kn*, and this can be resolved into two components, *km* and *kl* (each proportional to its length). Of these two, only *kl* is transmitted by the analyzer, *km* being absorbed. As the polarizer is rotated, the *kl* component changes its amplitude, becoming zero when the crystals are crossed at 90°. The colour is not changed, but the brightness is. The brightness of the transmitted light is proportional to the cosine of the angle between the polarizer and the analyzer.

When the polars of a microscope are crossed, no light is transmitted through it. When any isotropic substance, such as glass, is placed on the microscope stage (between the polarizer and the analyzer) it remains dark. If a crystal that is placed on the microscope stage also remains dark, that crystal belongs to the cubic system. A crystal of any other system, however, will allow some light to be transmitted because the *birefringence* (see below) re-orientates the plane of polarization and some light can then pass the analyzer.

Crystals affect the passage of light in some interesting and important ways. A ray of light entering a crystal of the cubic system is refracted just as it would be by glass and the refracted ray is in the plane of incidence; but all the other systems give rise to two refracted rays, at least one of which does not necessarily lie in the plane of incidence.

The *Nicol prism* is a device for polarizing and analyzing the directions of vibration of light. It consists of a crystal of calcite ($CaCO_3$) in a pure form called Iceland spar. Two slices of Iceland spar, cut normal to the optic axis, are cemented with Canada balsam. The incident light on the prism gives rise to two refracted rays, one of which is totally reflected by the balsam, the other transmitted. The transmitted ray is polarized. In a petrological microscope, one Nicol prism is used to polarize the light (the polarizer), another is used to analyze it (the analyzer). In modern petrological microscopes, the Nicol prism has been replaced by *polars* that are made of Polaroid™, but the effect is the same.

Optical activity

Some substances have the ability to rotate the plane of polarization of light passing through them. This is called optical activity. These substances, commonly organic, have asymmetric molecules. Crude oil is such a substance, and it is a strong argument for the organic origin of petroleum.

Pleochroism, dichroism and trichroism

Minerals in thin section that change colour, either hue or intensity, when rotated in polarized light are called *pleochroic* and the process is called *pleochroism*. Minerals with only one optic axis (those belonging to the hexagonal and tetragonal systems) show two characteristic colours. Those with two optic axes – the biaxial minerals, which belong to the orthorhombic, monoclinic, and anorthic systems – show three characteristic colours. Isotropic minerals are not pleochroic because the absorption of light is equal in all directions. Changes of colour or hue are due to unequal absorption of light of different frequencies (or wavelengths) in different orientations; and changes in intensity are due to loss of amplitude in different orientations.

Tourmaline (hexagonal, trigonal) is a uniaxial mineral, and we have seen that no polarized light is transmitted through crystals cut parallel to the optic axis when its axis is at 90° to the direction of polarization. In other orientations light is partly absorbed. Basal sections of uniaxial minerals do not show pleochroism because absorption of light of all wavelengths is equal in all directions. *Pleochroic haloes*, due to small inclusions of a radioactive contaminant, are common in tourmaline. Biotite (monoclinic, pseudo-hexagonal; biaxial) is a common pleochroic mineral. In sections other than basal, its colour changes from a dark brown when the cleavage is parallel to the plane of polarization, to yellow.

Specialists may use the term dichroism for uniaxial minerals, because these show two characteristic colours. Similarly, trichroism may be used for biaxial minerals, which show three characteristic colours.

Birefringence

Anisotropic crystals are found to have one index of refraction for light linearly polarized in one direction, and another for light linearly polarized in another direction. This is birefringence, and the crystal is said to be birefringent. Such crystals usually consist of non-spherical molecules with their long axes aligned and the speed of light is different in the two directions.

Calcite is strongly birefringent (which is why it was used in the Nicol prism). One ray is polarized in the plane of symmetry of the crystal (the *ordinary* ray, ω) and obeys the ordinary laws of refraction, the other normal to it (the *extraordinary* ray, ε) does not obey the ordinary laws of refraction. The refractive index of the ordinary ray, ω, of calcite is 1·658 while that of the extraordinary ray, ε, is 1·486. For quartz, the refractive index of the ordinary ray is 1·544 and of the extraordinary ray, 1·553.

All crystals of the hexagonal and tetragonal systems transmit light as two plane-polarized rays within the crystal, an ordinary ray and an extraordinary ray. They all have one direction in which there is no birefringence, and this is called the *optic axis.*

All crystals of the orthorhombic, monoclinic, and anorthic systems also show birefringence, but light is transmitted as two extraordinary rays. Crystals of these systems have another peculiarity: they show two axes that have most of the features of the optic axis of uniaxial crystals, and for this reason, they are called *biaxial.*

Luminescence: fluorescence and phosphorescence

Some substances, such as fluorite (CaF_2), have the property that light of one colour falling on it is absorbed and light of another colour is emitted. This is fluorescence. Light of higher frequency (shorter λ) is absorbed, and light of lower frequency (longer λ) is emitted. Equally, fluorescence is the absorption of a quantum of a certain energy, and the emission of one of lower energy (as one would expect).

The fluorescent lamp has an ultra-violet source inside the tube that excites light in the visible spectrum. Fluorite (or fluorspar, CaF_2) fluoresces in ultraviolet light, which is where the name fluorescence comes from. Some organic substances fluoresce in ultra-violet light (crude oil and kerogens, for example) and so can be recognized readily in samples. A organic dye called fluorescein is used as a water tracer because it fluoresces quite strongly at very small concentrations. It was used over 100 years ago to trace water that entered sink-holes where the Danube crosses the Malm limestone outcrop between Immendingen and Möhringen to the great spring at Aach, from which it flows to the Rhine (see Chapman 1981: 98).

Phosphorescence is a related property – the emission of light in the visible spectrum *after* being exposed to electromagnetic radiation (sunlight, heat, electricity, X-rays), or friction. A diamond, for example, after exposure to sunlight, is seen to glow in the dark.It glows more brilliantly after exposure to X-rays. Again, the emitted rays have a lower frequency (longer wavelength) than the absorbed. Quartz crystals glow in the dark after rubbing them together.

Diffraction

If you shine a small light on the edge of an opaque disc above a piece of paper, the shadow is not sharp. Some light rays passing near the edge are apparently bent. If you can examine the edge of the shadow, you will find that it is not just a diffused light, but consists of dark and light bands. This is diffraction. Diffraction is also the breaking up of light into dark and light patterns, or colours, when it passes through a grating or discontinuous solid (such as a crystal) in which the dimensions of the obstacles are not very much larger than the wavelength of the light.

The theory (due to Fresnel, and later to Huygens) is that each element of a wavefront acts as a source of vibration itself, sending out secondary waves. When a wavefront reaches an obstacle, some of these secondary sources of vibration are destroyed. The remaining sources – some in phase and so reinforcing each other, others out of phase – interfere with each other, giving the diffraction pattern of alternating light and dark bands. The width of these bands is proportional to the wavelength, so red light has wider bands than blue. White light leads to superimposed bands and a spectrum results.

Mother-of-pearl owes its colouring to diffraction of ordinary light due to very fine striations on the shell.

Diffraction gratings

Monochromatic light (i.e. of a single wavelength) passing through a transparent plate with many fine lines ruled onto it takes several paths:

- straight ahead as if there were no lines, and
- a scattering of paths as if the lines were the source of the radiation.

In Figure 3.5, the diffracted ray or beam travels farther than that above by a distance $a = d\sin\phi$, where d is the distance between rays or rulings. But the beams only reinforce each other when a is an exact multiple of the wavelength λ. So the *grating equation* is

$$n\lambda = d\sin\varphi \qquad (3.5)$$

where n is an integer and φ is the angle of diffraction. There will also be a symmetrical pattern diffracted upwards, and there is another solution for the supplement of φ, $180° - \varphi$. Note that if $d < \lambda$, Equation 3.5 can only have the solution $n = 0$, that is, $\varphi = 0$ or $180°$. This is not quite what it seems to be. It is not that all the radiation goes straight through: there is *new* radiation going in the same direction, *and* in the opposite direction.

Polychromatic light (many colours, many wavelengths) will fan out into spectra on either side of the normal to the grating (Fig. 3.6). If we have a grating with

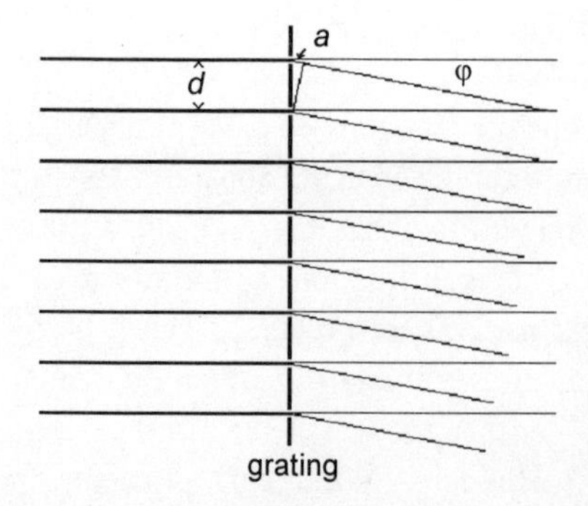

Figure 3.5 Diffraction of monochromatic light through a grating.
There is a symmetrical set of beams diffracted upwards.

lines at 5 μm spacing, a first-order radiation of 400 φm (just visible violet) will be diffracted through 4·59° (n = 1 in Eq. 3.5), as will second-order radiation of 200 nm (n = 2 in Eq. 3.5), third-order radiation of 133 nm, fourth-order radiation of 100 nm, and so on. At the other end of the visible spectrum, radiation of 750 nm will be diffracted through 8·63°, as will 375 nm, 250 nm, etc.

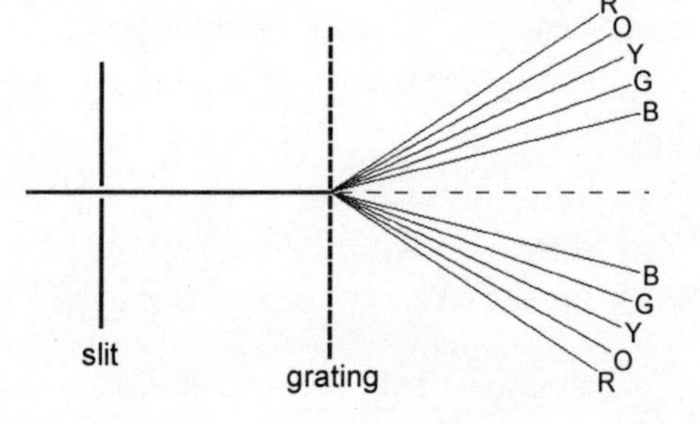

Figure 3.6 Diffraction of polychromatic light through a grating (diagrammatic).

This overlap of wavelengths does not lead to practical difficulty. If the observation is visual, all the second-order and above radiation is outside the visible spectrum. If the observation is photographic, emulsions can be chosen to detect specific ranges of wavelength (within limits).

Stereoscopy

We see things in three dimensions because our brain has the capacity to fuse the image each eye receives into one, each eye seeing a slightly different image from its slightly different position. Use is made of this in binocular microscopes and in the making of maps from aerial photographs. Photographs are taken sequentially from an aircraft flying at a constant height with the camera pointing vertically downwards, with about 60 per cent overlap between photographs. Adjacent pairs can be viewed stereoscopically through a stereoscope (see Fig. 3.7), and a three-

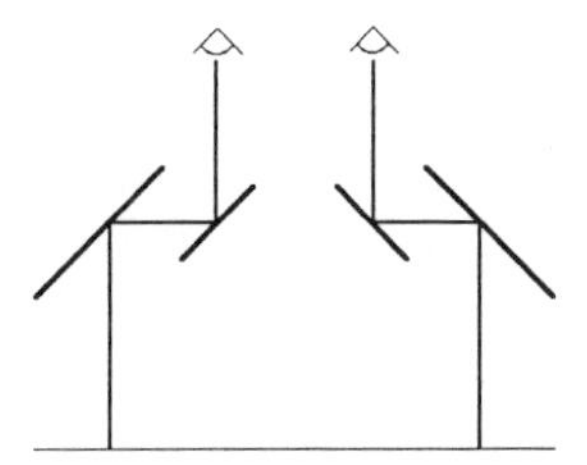

Figure 3.7 The mirrors in a stereoscope.

dimensional image of the common ground in the two photographs is seen. It is as if one eye was in the position from which the first photograph was taken, the other in the next. Topography seems exaggerated. Knowing the height of the aircraft above the ground, the focal length of the lens of the camera, and the horizontal distance between the *principal points* (the intersection of the optic axis of the camera with the ground and with the film), quite accurate maps can be drawn including contours and the heights of hills and buildings. The top of a tall factory chimney, for example, will be closer together in the two photographs than the bottom – that is, the distance between them on the table when viewed stereoscopically. An area is covered by *runs* and the runs must also overlap for a good map to be drawn.

In Figure 3.8, see if you can fuse the two images into a single stereoscopic image. If necessary, try placing a card vertically between the two circles to encourage each eye to look at one circle only, concentrating on the dark circles. You will see a right pyramid below the thick circle, and a thin circle above the thick circle, and various triangles and points.

Air photographs differ from maps in two main respects. A map is of constant scale, and the position of all points is the vertical projection of each point onto a horizontal surface. Looking at the geometry of air photographs in a little more

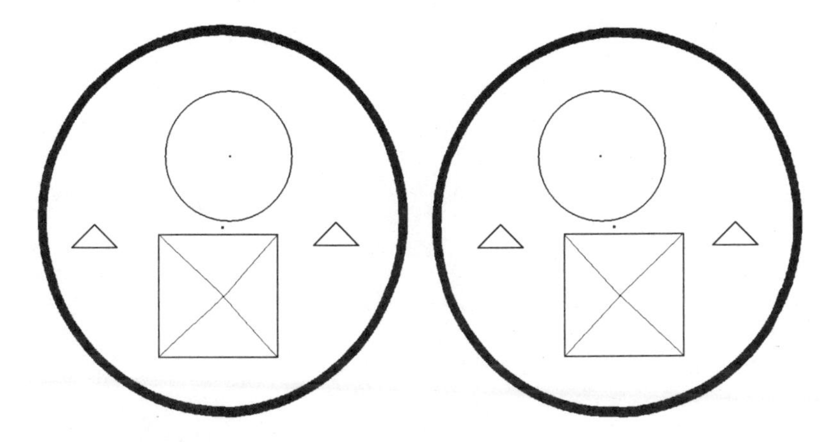

Figure 3.8 These figures can be viewed stereoscopically.

Concentrate on the black rings, one to each eye. It may help to separate the images with a card in front of your nose. Note that the differences of parallax create the impression of three dimensions.

detail we see that the scale of the photographs is the focal length divided by the height above ground (always in consistent units). This means that over hilly ground the scale changes according to elevation. The other difference is that the camera "looks" from a single point, so there is both distortion radially outwards from the principal point and also distortion due to the angle of view. For example, a small steep hill will appear farther from the principal point in a single photograph than it really is, and a valley nearer; and the camera may not see down a steep gorge near the edge of the photograph. The radial distortion can be corrected by triangulation if a feature appears on three consecutive photographs (as it should with 60 per cent overlap).

The line joining the principal points of the photographs of a run is called the *base line*. The change of apparent position of an object on account of a change of viewpoint is called *parallax*, which is measured parallel to the base line. In the figure, the parallax of the base of the pyramid is 55·5mm, of the top, 54·3mm (reproduction may have altered these figures) and it is the difference of parallax that creates the stereoscopic image. If there is no difference of parallax between points in the two images, the points appear to be at the same elevation.

For a more detailed, but still concise, description of basic photogeological techniques, see Allum (1966).

4
ATOMIC STRUCTURE

It used to be thought that the atom was the smallest indivisible particle of matter. We now know, and have known since radioactivity was discovered late in the 19th century, that atoms consist of various particles that carry a positive, a negative or no electric charge. It is important to have at least an outline knowledge of atomic structure in order to understand radioactivity and isotopes (not to mention geochemistry). The atom is sometimes called a *nuclide* when the nucleus is the focus of attention.

The main particles are:

- **neutron** Part of the nucleus of an atom, carrying no electric charge. Its mass is $1{\cdot}674\,7 \times 10^{-27}$ kg, rather larger than that of a hydrogen atom with a mass of $1{\cdot}673\,3 \times 10^{-27}$ kg. The neutron number, N, is the number of neutrons in the nucleus.
- **proton** Part of the nucleus of an atom, carrying a single positive charge. Its mass is $1{\cdot}672\,4 \times 10^{-27}$ kg, about the same as that of a neutron. The *atomic number*, Z, is the number of protons in the nucleus.
- **electron** Very small particles, carrying a single negative charge of electricity. They are in clouds or swarms around the nucleus. The mass of an electron is $910{\cdot}9 \times 10^{-33}$ kg.

Other particles are:

- **α-particles** Really four particles, consisting of two neutrons and two protons. An α-particle may be emitted from the nucleus during radioactive decay. It is identical to the helium atom's nucleus.
- **β-particles** Identical to the electron, but may be positively (β^+) or negatively charged (β^-). They may be emitted from an atom's nucleus during radioactive decay. When positively charged, they are called positrons.
- **positron** Particles of the same mass as electrons but with a positive charge of electricity. A positron is thus the *antiparticle* of the electron.
- **neutrinos** Particles, probably with no rest mass[1], carrying no electric charge.

- **photon** Discrete packets of energy (quanta) of electromagnetic radiation. The amount of energy depends on the frequency of the radiation and covers the full range of electromagnetic frequencies (p. 53). They have no charge, and no rest mass. They travel at the speed of light.

As a matter of interest, there is another aspect of atomic structure that is not intuitively obvious: *Heisenberg's uncertainty principle* or *indeterminacy principle*. This states that the position and speed of a particle cannot both be measured with precision at the same time. Strictly, it applies to all objects, but it is only significant with subatomic masses. The basic reasoning is that any observation involves a change of state – either the emission of photons if it is to be seen, or the bombardment by photons if it is to be illuminated. (See Feynman et al. 1963, vol. 1, 6–10.)

An atom is very small[2]: about 100×10^{-12} m, or 100 pm. Each atom consists of a positively charged *nucleus* surrounded by a cloud of negatively charged *electrons*. The nucleus consists of two types of particles: *neutrons* and *protons*. Atoms as a whole have, in general, no charge so the number of the electrons must equal the number of the protons. The simplest atom of all, hydrogen, consists of a nucleus of one proton accompanied by a single electron. The nucleus is very, very small compared with the size of the atom, the volume ratio being about 10^{-15}, in spite of contributing most of its mass.

Atoms have mass, and so weight, but common usage makes no distinction between the two, and *weight* and *mass* are used interchangeably. *Atomic weights* were originally expressed relative to hydrogen until it was found that hydrogen had three atoms that slightly differed from each other. These *isotopes* (to be explained below) made hydrogen unsuitable as a standard, so the unit definition was changed to 1/12 of the mass of the common carbon atom that has 12 neutrons and protons in its nucleus, the atomic weight of which is 12·011. The mass of an atom of carbon–12 is $19{\cdot}924 \times 10^{-27}$ kg, so one *atomic mass unit* (a.m.u.) is $1{\cdot}660\,3 \times 10^{-27}$ kg.

The electron "cloud" around a nucleus must not be thought of as a minute model of the Solar System, with electrons in orbit around the nucleus. Newtonian mechanics are totally inadequate in describing the motion of electrons around their nucleus. The fundamental force between the nucleus and its electrons is electromagnetic. Unlike charges attract each other: like charges repel. So the electrons are attracted to the nucleus but repel each other. Moving charges generate a magnetic field.

Particles have *energy*, and the unit is the *electron volt*, eV, $[ML^2T^{-2}]$. This is

1. Rest mass is the term m_0 in the equation $m = m_0/\sqrt{1-(v/c)^2}$ where v is the speed of the particle and c is the speed of light.
2. The old unit, the angstrom, named after the Norwegian physicist Anders Jonas Ångström, was a good practical unit because most atoms are about 1 Å in diameter (100×10^{-12} m).

the energy acquired by a particle carrying a single charge when acted on by a potential difference of 1 V; 1 eV = $160{\cdot}210 \times 10^{-21}$ joules (J). Be careful with units and dimensions here because the symbol is to be seen as a single symbol: eV, not $e \times V$. An electric charge in coulombs [A s] acted on by a volt [$W A^{-1}$] gives an energy in watt-seconds which equals joules [$W = J s^{-1}$]. The charge on an electron (*e*) is a fundamental and natural unit of electricity. Its value is $160{\cdot}210 \times 10^{-21}$ coulombs (C), and all non-zero charges are a whole multiple of this.

The number of protons is called the *atomic number* (*Z*). Elements are placed in the periodic table according to their atomic number because an element is determined by the charge of the protons in the nucleus, not by the neutrons. Neutrons and protons are much more massive than electrons (by factors of 1836·1 and 1838·63 respectively), and the total *number* of neutrons (*N*) and protons (*Z*) is called the *atomic mass number* (*A*). This is a whole number that is very nearly equal to the mass of the atom because of the small mass of the electrons. The full notation is given by prefixes to the abbreviation of the element, the atomic number written subscript; and the atomic mass number written superscript $^{A}_{Z}$. For example, $^{12}_{6}C$ means that there are 6 protons, and 12 protons and neutrons (so, 6 neutrons) in the nucleus. The symbol is often abbreviated by leaving out the atomic number, e.g. ^{12}C. In older works you may find the superscript prefix after the element abbreviation, as C^{14}.

Some atoms of simple elements have differing numbers of neutrons. Hydrogen is found in three forms, $^{1}_{1}H$, $^{2}_{1}H$, and $^{3}_{1}H$, which we call hydrogen, deuterium and tritium. These are *isotopes* of hydrogen (see also isotopes and age-dating in Ch. 5). Another example is carbon, which also has three isotopes: $^{12}_{6}C$, $^{13}_{6}C$, and $^{14}_{6}C$. Isotopes of an element have the same general chemical properties (because protons, not neutrons, determine this) but differ slightly in mass and in some physical properties. They occupy the same place in the periodic table because they have the same atomic number.

Isotopes are of two sorts: stable and unstable. Certain combinations of neutrons and protons are stable, others unstable. The unstable isotopes change spontaneously to a more stable form by the emission or capture of particles. These *unstable isotopes* are also called *radioisotopes*. It turns out that stable isotopes are those in which *Z* and *N* are nearly equal (or *Z* is about half *A*) – a small minority of the isotopes. Tritium, $^{3}_{1}H$, is an unstable isotope of hydrogen. The *decay* of an unstable isotope also involves the emission of *quanta* of energy. $^{14}_{6}C$ is a radioisotope and it will be discussed later.

There are several different forms of radioactive decay. The summary of the important ones that follows is extracted largely from Faure (1986), to which reference could be made for more detailed discussion:

- *β-decay*. The prime event is the emission of a β-particle from the nucleus, along with some neutrinos and often γ-rays. The atomic number, *Z*, is increased by one, and the neutron number, *N*, is reduced by one. This changes the element. For example, potassium becomes calcium with the emission of a β-particle

$${}^{40}_{19}\mathrm{K} \rightarrow {}^{40}_{20}\mathrm{Ca} + \beta^-$$

This can be interpreted as the transformation of a neutron into a proton and an electron.

- *Positron decay*. A positron is a positively charged electron from the nucleus and when one is emitted the atomic number, *Z*, is decreased by one, and the neutron number, *N*, is increased by one. This changes the element. For example, ${}^{18}_{9}\mathrm{F} \rightarrow {}^{18}_{8}\mathrm{O} + \beta^+$. This can be interpreted as the transformation of a proton in the nucleus into a neutron, a positron and a neutrino.
- *Electron capture*. When an electron from the "cloud" is captured by the nucleus (usually an electron nearer the nucleus), the nucleus emits a neutrino. The atomic number, *Z*, is decreased by one, and the neutron number, N, is increased by one. This changes the element. For example, ${}^{40}_{19}\mathrm{K} + e^- \rightarrow {}^{40}_{18}\mathrm{Ar}$. This can be interpreted as a reaction between an electron and a proton in the nucleus to form a neutron and a neutrino.
- *Branched decay*. Some nuclides decay by more than one process, such as ${}^{40}_{19}\mathrm{K}$ to both ${}^{40}_{18}\mathrm{Ar}$ and ${}^{40}_{20}\mathrm{Ca}$, which will be discussed later.
- α-decay. The emission of α-particles from atoms having atomic numbers, *Z*, greater than 58, and from a few with small atomic numbers. This reduces both the atomic number and the neutron number by two and so changes the element. For example, ${}^{238}_{92}\mathrm{U} \rightarrow {}^{234}_{90}\mathrm{Th} + {}^{4}_{2}\mathrm{He}$.
- *Ionization* is a process that converts neutral atoms or molecules to charged atoms or molecules. It has slightly different meanings in physics and chemistry. Ionization can occur by collision, ejecting an electron from a neutral atom if the energy is sufficient. It may occur at very high temperatures, if the temperature is high enough to excite the atoms sufficiently to part with an electron. In aqueous solutions, the molecules dissociate into cations (+) and anions (–), which can be manipulated by electric currents.

Cosmic rays ionize the atmosphere to some extent. So the *ionosphere* is the part of the Earth's outer atmosphere that contains ions and free electrons – a zone that affects radio propagation. X-rays and γ-rays ionize the material they pass through to some extent.

5
ELECTROMAGNETIC RADIATION

All radiation, whether we are considering the warmth given out from a fire, the light from a lamp, or the X-rays in a hospital, is electromagnetic by nature and propagated at the speed of light. Electromagnetic radiation behaves in some respects as waves, in others as particles. We categorize radiation by its frequency of oscillation (see Table 5.1) and/or by its wavelength. It must be remembered that there is no sharp dividing line between the categories; X-rays grade into γ-rays, and their properties also merge, and radar waves also heat, much as the microwave oven does. The subdivision is based on general characteristics, such as the visible part of the spectrum or the frequencies suitable for radio transmission.

Frequency may be regarded as more fundamental than wavelength because a particle can oscillate with a frequency even if it does not behave as a wave; and frequency is independent of the medium through which the radiation travels, whereas wavelength is not (as we saw when considering the refraction of light).

Table 5.1 Electromagnetic radiation.

	Approximate values	
	ν (Hz)	λ (m)
γ-rays		
	3×10^{21}	100×10^{-15}
X-rays		
	300×10^{15}	10^{-9}
ultraviolet (UV)		
	800×10^{12}	370×10^{-9}
visible light		
	400×10^{12}	750×10^{-9}
infrared (IR)		
	10^{12}	300×10^{-6}
microwaves		
	10^{9}	300×10^{-3}
radio waves		
	10×10^{3}	30×10^{3}

Cosmic radiation

Cosmic radiation, which was discovered early in this century, consists of very high frequency, high energy X-rays (> γ-rays) together with very high energy particles. The general level was found to be roughly constant over 24 h, so clearly the radiation does not come from the Sun. The intensity was found, by apparatus flown in balloons, to increase up to a height of about 12 km, then to decrease; so some cosmic radiation, or its effect, is clearly generated within the atmosphere. Cosmic radiation is much more intense in the polar regions than the equatorial, so clearly the particles are charged (the equatorial regions being protected by the magnetic field's being approximately tangential to the Earth's surface near the Equator).

γ-rays

γ-rays are high-energy radiation from radioactivity with frequencies greater than 3×10^{21} Hz (wavelengths shorter than about 100×10^{-15} m or 100 fm), and their significant property is that they are not affected by magnetic fields, so they are not charged particles. α and β-particles are affected by magnetic fields.

γ-rays are emitted spontaneously from the nucleus of some unstable isotopes. They are gradually absorbed when passing through solids, but can penetrate some millimetres of steel, some centimetres of concrete, and rather more centimetres of sedimentary rock. The main isotopes yielding γ-rays are of the uranium and thorium families, and ^{40}K (potassium–40). Potassium-40 occurs in the micas and K-feldspar, and clay minerals by cation exchange. Mudrocks or mudstones have a level of γ-radiation that is significantly higher than that of sandstones and most carbonates. The natural γ-radiation is therefore a useful borehole-logging parameter.

When γ-rays are absorbed by a nucleus, neutrons and α-particles are emitted. γ-rays enhance catalytic activity. Most clay minerals act as catalysts in petroleum-chemical reactions, and most emit γ-rays spontaneously, and this may be significant in the generation of petroleum.

X-rays

X-rays were discovered in 1895 by Wilhelm Röntgen when he found fluorescence in a cathode tube when electrons were beamed from the cathode (negative electrode). This is roughly how they are produced today for medical and dental purposes. The X stands for "unknown". X-rays are emitted from an excited atom, varying in intensity and wavelength from element to element. They are produced in a vacuum tube by accelerating electrons in a high-voltage field from the cathode to collision with the anode, where the X-rays are generated. Like γ-rays, they are not deflected by a magnetic field, so they are not charged particles.

X-rays have frequencies between about 100×10^{15} Hz and about 3×10^{21} Hz (wavelengths between about 10^{-9} and about 100×10^{-15} m, shorter than ultra-violet and so of a higher frequency). Like other electromagnetic radiation, they show wave characteristics of interference, diffraction and polarization. They show particle characteristics in their scattering, and the fluorescence they excite in some materials (a property used in analysis, as we shall see). They have varying penetrative power, and affect photographic emulsions. When X-rays pass through material, the material itself becomes a source of X-rays and electrons. The secondary X-rays consist of scattered and fluorescent rays, the scattered being of about the same energy as the primary rays, the fluorescence being weaker, as you would expect. X-rays can be polarized by a block of carbon.

Light can be diffracted by passing it through a finely ruled grating. In 1912, Max Laue, a German physicist, thought of measuring the wavelength of X-rays by passing them through crystals and measuring the diffraction, replacing the grating by the lattice of the crystal. Soon X-rays were to be used to analyze the atomic lattice structure of crystals, and X-rays and crystallography have been closely related ever since.

X-rays also originate in space (as do infrared and radio waves), and such sources seem to lie close to the galactic plane (through the Milky Way) with few exceptions. Very few of these sources are identifiable with visible objects. The Sun is one.

X-ray diffraction (XRD)

X-rays, being electromagnetic radiation, can be diffracted like light. The grating equation, (Eq. 3.5), $n\lambda = d \sin \varphi$, applies; but since λ is of the order of 1 nm, d has to be smaller than is possible (or practicable) in a manufactured grating – say, 1 μm. Fortunately crystals commonly have the right spacing between the layers of atoms: sodium chloride, calcite, and gypsum, for example.

If the X-ray beam is inclined to the crystal (Fig. 5.1) at an angle θ, there will be a scattering from successive layers of atoms. Because the "reinforcing" parallel path lengths differ by $2d \sin\theta$, we have the *Bragg equation*:

$$n\lambda = 2d \sin\theta \qquad (5.1)$$

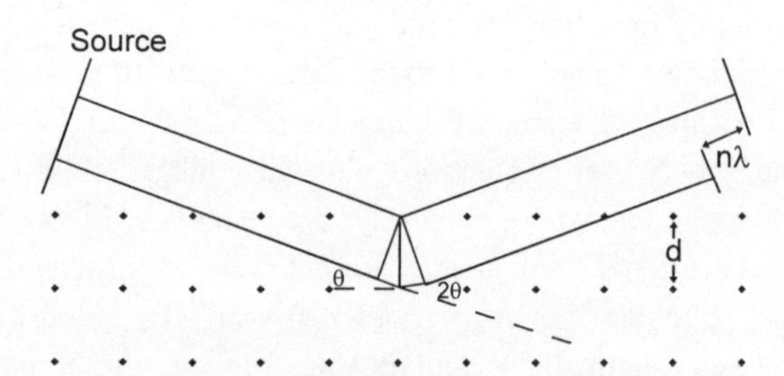

Figure 5.1 X-ray diffraction.
The lower path is longer than the upper by $2d\sin\theta$.

where θ is the angle of incidence, *d* is the distance between layers of crystal atoms, and *n* is an integer. Different angles diffract different wavelengths.

X-ray diffraction analysis is based on the assumption that no two crystals of different composition have identical atomic spacing, and that measurement through all possible values of θ will give an unique pattern of wavelengths. The Debye–Scherrer powder camera (Fig. 5.2) is used for simple solids that are fairly pure. The powder is embedded in a non-crystalline medium and it is assumed that random orientation of the particles will cover the full range of θ. Photographic film is in a circular holder. After exposure and development, the film has lines that are spaced according to θ, symmetrically arrayed around the point θ = 0°.

X-ray fluorescence (XRF) or X-ray emission spectroscopy

When metals and other massive samples with atomic numbers less than about 20 are irradiated with high-energy X-rays, characteristic X-rays of lesser energy are excited. These wavelengths are analyzed using XRD methods and a crystal grating of known spacing (an X-ray spectrometer, which is strictly analogous to the grating spectrometers of visible light). Different analyzing crystals are used for different ranges.

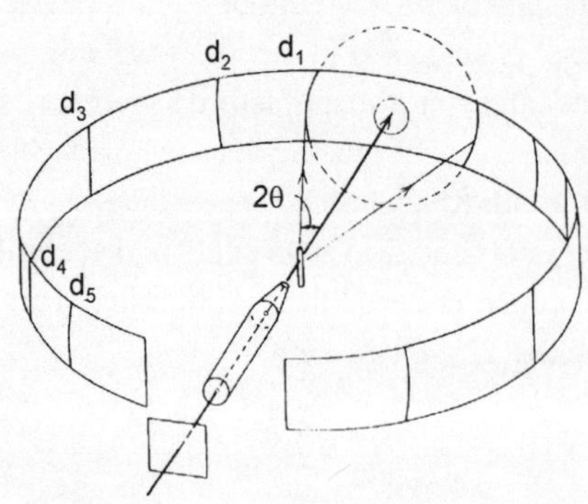

Figure 5.2 The Debye–Scherrer powder camera.
The powder embedded in a non-crystalline medium is orientated at random to X-ray path (courtesy Philips Nederland BV).

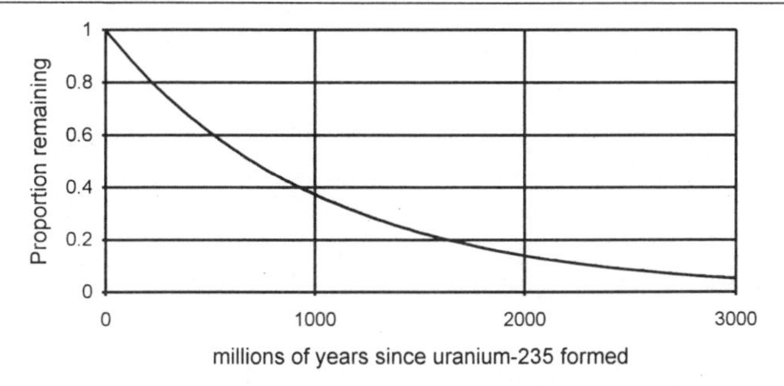

Figure 5.3 The decay of uranium-235 (half-life 704 million years).

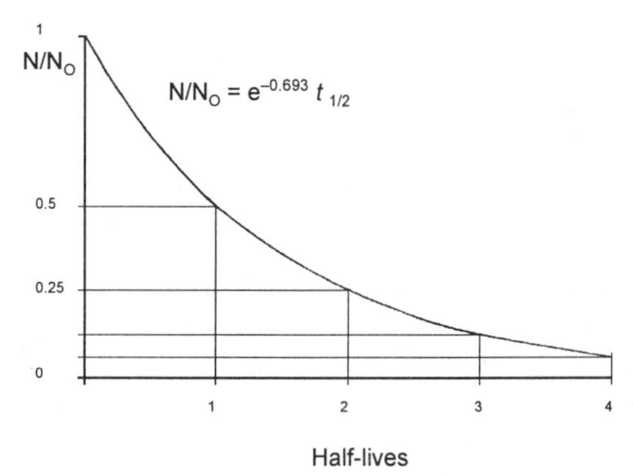

Figure 5.4 Radioactive decay in terms of half-lives.
N/N_0 is the proportion of parent nuclide remaining.

Isotopes and age-dating

Isotopes are species of the same chemical element that have different masses, as we have seen. They occupy the same position in the periodic table because it is the atomic number, *Z*, the number of protons, that determines the element. Each has a nucleus with the same number of protons (equal to *Z*), but varying numbers of neutrons. For example, the three isotopes of hydrogen have already been mentioned: hydrogen itself (atomic weight 1), deuterium (atomic weight 2) and tritium (atomic weight 3). They each have one proton in the nucleus but differ in the number of neutrons. Isotopes have the same basic chemical composition, but their physical properties may differ a little. In particular, they diffuse at different rates because diffusion depends on the mass of the diffusing species.

Unstable isotopes differ from stable isotopes in that the change from one to another is spontaneous. Radioactive substances change spontaneously from the *parent nuclide* to *daughter nuclides.* In 1896 Becquerel discovered that uranium salts affected photographic plates that were wrapped in paper, and it is now known that uranium isotopes change spontaneously to lead isotopes as follows:

$${}^{238}_{92}U \rightarrow {}^{206}_{82}Pb + 8\,{}^{4}_{2}He + 6\beta^{-}$$

$${}^{235}_{92}U \rightarrow {}^{207}_{82}Pb + 7\,{}^{4}_{2}He + 4\beta^{-}$$

Thorium also changes to an isotope of lead; and rubidium changes to an isotope of strontium by the loss of a β-particle:

$${}^{232}_{90}U \rightarrow {}^{208}_{82}Pb + 6\,{}^{4}_{2}He + 4\beta^{-}$$

$${}^{87}_{37}Rb \rightarrow {}^{87}_{38}Sr + \beta^{-}.$$

Potassium–40 decay is more complicated. It decays to both ${}^{40}_{18}Ar$ and to ${}^{40}_{20}Ca$. It changes in part to argon by a reverse process, *electron capture*, the addition of an electron:

$${}^{40}_{19}K + e^{-} \rightarrow {}^{40}_{18}Ar$$

but mostly to ${}^{40}_{20}Ca$ by β-particle emission.

Moreover, each radioisotope or radionuclide decays at a statistically constant rate that is proportional to the amount of parent substance remaining (Figs 5.3, 5.4). It takes a fixed period of time for the parent nuclide to convert half its atoms to a daughter nuclide; and the same time to reduce that number of atoms to half its number; and so on. This period, known as the *half-life* ($t_{½}$), is different for each parent nuclide. The half-lives[1] of those isotopes above are:

^{238}U $4{\cdot}47 \times 10^{9}$years
^{235}U 704×10^{6} years
^{232}Th $14{\cdot}0 \times 10^{9}$ years
^{87}Rb $48{\cdot}8 \times 10^{9}$ years
^{40}K $1{\cdot}4 \times 10^{9}$ years to ^{40}Ca
^{40}K $11{\cdot}9 \times 10^{9}$ years to ^{40}Ar
^{40}K $1{\cdot}25 \times 10^{9}$ years total.

The three isotopes of carbon have already been mentioned: ${}^{12}_{6}C$, ${}^{13}_{6}C$, and ${}^{14}_{6}C$. The first two are stable, nearly 99 per cent of carbon atoms being ${}^{12}C$. *Radiocarbon*, as ${}^{14}C$ is called, is created from nitrogen in the atmosphere by cosmic radiation. Fast neutrons are emitted from collisions with oxygen and nitrogen atoms, and these fast neutrons in turn collide with nuclei of ${}^{14}_{7}N$, dislodging a proton and leaving ${}^{14}_{6}C$. In due course, the isotope ${}^{14}_{6}C$ returns to ${}^{14}_{7}N$ through the loss of a negative β-particle: ${}^{14}_{6}C \rightarrow {}^{14}_{7}N + \beta^{-}$. All living things absorb some ${}^{14}C$, but cease to do so when they die. The half-life of radiocarbon is about 5730 years, and

1. Derived from the values of the decay constant, λ, given by Steiger & Jäger (1977). $t_{½} = (\ln 0{\cdot}5)/{-\lambda}$.

dating of organic material from about 500 to 30 000 years is possible. Unfortunately, the natural balance of carbon isotopes is being disturbed by CO_2 from the burning of fossil fuels, which lost their ^{14}C long ago.

The law of radioactive decay was discovered by Rutherford & Soddy in 1902. They found that the activity of a radioactive substance decreased with time according to the relationship

$$N = N_0 e^{-\lambda t}, \tag{5.2a}$$

where N_0 is the original number of radioactive atoms and N is the number after time t (Fig. 5.4). The material constant λ is known nowadays as the *decay constant*, and its inverse ($1/\lambda = \tau$) is known as the *mean life*. Differentiating, $\mathrm{d}N/\mathrm{d}t = -\lambda N_0 e^{-\lambda t} = -\lambda N$, and the rate of decrease in the number of radioactive atoms is proportional to their number. (Note that this would have been a reasonable postulate, and integration would have led to the same equation.) No known agency, physical or chemical, can control the rate of decay except bombardment by other particles. It is purely a matter of chance.

Equation 5.2a leads to an expression in terms of half-life, $t_{½}$, without λ explicitly:

$$t/t_{½} = -(\ln N - \ln N_0)/\ln 0{\cdot}5, \tag{5.2b}$$

but λ is hidden in there in $t_{½} = \ln(0{\cdot}5)/{-\lambda}$. Measurement of the concentrations of parent and daughter nuclides in minerals and rocks by chemical or physical means enables geochemists to determine the age of the minerals or rocks. The age represents the time at which the mineral or rock became a closed chemical system as regards the nuclides concerned.

There are some important assumptions made in the determination of so-called *absolute* dates or ages from isotope analysis. These are:

(a) that the daughter did not exist at time zero when the parent nuclide was formed,

(b) that the decay constant is in fact a constant, and is reasonably accurately known,

(c) that nothing can significantly alter the decay rate, and

(d) that parent and daughter can be measured accurately, and that their quantities have not been altered by leakage or addition.

Assumption (a) is implicit in the equations. Assumptions (b) and (c) are supported by theory and experiment. Assumption (d) is necessary, and consistent results tend to support it. Special care has to be taken with the measurement of gases, as in the potassium-to-argon decay, and any leakage is usually detected by inconsistent results when checked by other methods.

Does a half-life of $48{\cdot}8 \times 10^9$ years for ^{87}Rb mean that the universe is at least that old? . . . or will last that long?

6
HEAT & HEAT FLOW

The noun *heat* has several different meanings:

- the quality of being hot, as in heat wave,
- the perception of this quality, and
- the quantification of this quality, e.g. *temperature*.

Heat is a form of energy (hence the dimensions) and it is conserved. Bodies have an internal energy due to their heat, and heat cannot flow from a cold body to a hot body without an input of energy. Temperature [$L^2 T^{-2}$ in dynamical units or θ in thermal units] is a measure of the hotness of an object or material, but it is not a measure of the amount of heat [$ML^2 T^{-2}$] in the object or material. This must depend on the size and the nature of the object. It takes more heat to raise the temperature of 2 litres of water from 20 to 100°C than it does for 1 litre; and if you add 1 litre at 100°C to the 2 litres at 100°C you have 3 litres at 100°C. Temperature is a measurement of heat per unit of mass. Table 6.1 lists the quantities we are mostly concerned with in geology.

Table 6.1 Heat: common symbols, units and dimensions.

	Symbol	Unit	Dynamical	Thermal
Quantity of heat	Q	J	ML^2T^{-2}	
Temperature	T, Θ, θ	K, °C		θ
Heat flow rate	Φ	W	ML^2T^{-3}	
Density of heat flow rate	q	$\mathrm{W\,m^{-2}}$	MT^{-3}	
Heat capacity	C	$\mathrm{J\,K^{-1}}$		$ML^2T^{-2}\theta^{-1}$
Specific heat capacity	c	$\mathrm{J\,kg^{-1}\,K^{-1}}$		$L^2T^{-2}\theta^{-1}$
Thermal conductivity	λ, k	$\mathrm{W\,m^{-1}\,K^{-1}}$		$MLT^{-3}\theta^{-1}$
Thermal gradient	$\Delta\theta/l$	$\mathrm{°C\,m^{-1}}$		$L^{-1}\theta$

Heat can be transferred. If you sit in front of a fire, you can feel its heat. A steak is heated under a grill. This is *radiated* heat, and the process is *radiation* (yes, some radiation is beneficial!). As you solder a wire, the heat moves up the wire towards your fingers and they feel heat. This is *conducted* heat and the process is called *conduction.* If you watch a saucepan of water when it is nearly boiling, you see currents in it tending to bring the hot water from the bottom to the top. This is *convected* heat, and the process is called *convection.* Convection is essentially gravitational because warm water is less dense than cold water; it rises and displaces the cooler water above, which sinks and is duly heated, to rise again. (Note: once steam bubbles form, the process changes because friction around the rising steam bubbles generates a sympathetic flow.) Radiated heat consists of packets of electromagnetic energy that are converted to heat in the substance absorbing the radiation. Heat is conducted through materials at a molecular level. The kinetic theory asserts that molecules in solids, liquids and gases are in constant motion and behave as perfectly elastic particles with their mean kinetic energy proportional to their temperature. The greater energy of the hotter molecules is transferred to neighbouring molecules, and so spreads throughout the conductive material. It is a process of *diffusion.*

Heat can be generated by various processes. *Friction* generates heat (as in the brakes of your car, or in your hands when sliding down a rope). Hammering a metal warms it. Some organic processes generate heat, particularly decomposition (as in a pile of grass cuttings). Some chemical reactions give out heat (exothermic) others absorb heat (endothermic). A liquid can lose heat by evaporation.

The laws of thermodynamics were formulated from practical observation (of steam engines especially) as well as experiment. The first law concerns the conservation of energy, such as the dissipation of energy by friction with the generation of heat. Joule found that the amount of heat produced in a mechanical device for churning water was proportional to the work done in churning it. This led in 1847 to the principle of conservation of energy, propounded by Joule in Britain and by Helmhotz in Germany. The second law concerns heat and temperature. If you put two bodies of different temperature together, such as a cold egg in a saucepan of warm water, heat will flow from the hotter to the cooler until both are at the same temperature. The egg does not become colder and the water hotter. The hotter object is said to have greater *entropy* than the colder. Entropy is a measure of the disorder in a system. The third law says that entropy is zero at absolute zero temperature. The laws of thermodynamics have been stated in various ways. Feynman et al. (1963: vol. 1, 44-13-, Table 44-1) summarized them as follows:

First Law:
Heat put into a system + Work done on a system = Increase in internal energy of the system:

$$\mathrm{d}Q+\mathrm{d}W=\mathrm{d}U.$$

Second Law:
A process whose *only* net result is to take heat from a reservoir and convert it to work is impossible.

No heat engine taking heat Q_1 from T_1 and delivering heat Q_2 at T_2 can do more work than a reversible engine, for which

$$W = Q_1 - Q_2 = Q_1\left(\frac{T_1 - T_2}{T_1}\right)$$

The entropy of a system is defined this way:

–If heat ΔQ is added reversibly to a system at temperature T, the increase in entropy of the system is $\Delta S = \Delta Q/T$.

–At $T = 0$, $S = 0$ *(Third Law).*

In a *reversible change*, the total entropy of all parts of the system (including reservoirs) does not change.

In *irreversible change*, the total entropy of the system always increases."

Heat flow

The interior of the Earth is very hot; the surface is sufficiently cool for us to live on it comfortably. There is a continuous transfer of heat from the interior of the Earth to the surface, where it is dissipated into the atmosphere by conduction and convection in air and water. Some of this heat is due to radioactivity, some to the deep-seated primeval heat of the Earth arising from gravitational processes during the early stages of the Earth's formation. The rate at which heat is transferred by conduction near the surface (where convection cannot be important because of the rigidity of the crust) depends on the *thermal conductivity* of the material conducting the heat, and its specific heat capacity; and these give rise to a *geothermal gradient.* Heat flow is a process of diffusion, and a finite time is require for a stable thermal gradient to be achieved. The thermal conductivity of rocks is very small. Seasonal changes of temperature are only felt within metres of the surface of the Earth, which is why some people like to live in caves. Even relatively shallow boreholes produce water at a virtually constant temperature. It has been calculated that we would not feel heat from an original source below a depth of a few hundred kilometres, created with the world, because it would not yet have reached the surface.

The quantity of heat, Q $[ML^2T^{-2}]$, that flows through an area A and thickness l in time t is given by

$$Q = -k\,t(A/l)\,\Delta\theta, \qquad (6.1a)$$

where k is the *thermal conductivity* and $\Delta\theta$ is the difference of temperature across thickness l. We are usually more interested in heat per unit of time and per unit area, so

$$q = -Q/At = k\Delta\theta/l. \qquad (6.1b)$$

The equation is properly written with a minus sign before the k to indicate the direction of flow towards smaller energies. The larger the thermal conductivity, the smaller the geothermal gradient. The dimensions of k are $(ML^2T^{-2})(T^{-1})(L^{-2})(L) = MLT^{-3}$, per K or °C; in units, $\mathrm{W\,m^{-1}K^{-1}}$. For all rocks, the thermal conductivity is very small, but different lithologies have different conductivities. The *heat flow unit* is $\mathrm{mW\,m^{-2}}$; but the older literature has units of $\mathrm{\mu cal\,cm^{-2}\,s^{-1}}$, which is equal to $41{\cdot}87\,\mathrm{mW\,m^{-2}}$.

Within the Earth there are surfaces of equal temperature, called isothermal surfaces. Heat flow is normal to these surfaces, just as water flow is normal to surfaces of equal energy (equipotential surfaces).

The geothermal gradient (which is the thermal gradient vertically into the Earth) in a homogeneous rock is

$$\Delta\theta/l = Q/(ktA) = q/k. \qquad [L^{-1}\theta] \qquad (6.1c)$$

The product of the geothermal gradient and the thermal conductivity is the heat flow, q (in $\mathrm{W\,m^{-2}}$), more properly called the density of heat flow rate. But a sequence is made up of rocks of different conductivities, and so *if possible* we should take them all into account, with their thicknesses:

$$q = \Delta\theta/[(l_1/k_1)+(l_2/k_2)+\ldots+(l_n/k_n)]. \qquad (6.2)$$

Typically, sedimentary rocks have thermal conductivities between 1 and $3\,\mathrm{W\,m^{-1}K^{-1}}$ (see Clark 1966: 459–82) while for frozen soil with 20 per cent moisture content it is about $0{\cdot}2\,\mathrm{W\,m^{-1}K^{-1}}$.

Permafrost and ice-caps

There is a special case of interest to do with ice. In the polar regions there are both ice-caps and, in places, thick sequences of rocks in which the pore water remains frozen all year below a depth of a metre or so. Such frozen rocks are called permafrost, and in some areas of North America the permafrost reaches thicknesses of 600–700 m, and in Siberia even double that figure. Permafrost is not uniformly thick: its greatest thickness seems to be in young sedimentary basins, and it is absent under rivers, lakes that do not freeze entirely during the winter, and under the polar seas. Its areal limit corresponds quite closely to the area with mean annual temperatures below 0°C, with the variations noted above. The question is, what is the geothermal gradient in regions of thick permafrost?

First, it is clear that at the base of the stable permafrost in a sedimentary sequence the temperature is close to 0°C, and it decreases upwards from there. The thermal gradient in permafrost is about 0·04°C/100 m or 0·4°C km^{-1} (but for computational purposes, thermal gradients must be expressed in °C m^{-1}). So, in Prudhoe Bay on the north slope of Alaska where the permafrost is 700 m thick, a mean annual surface temperature of about –28°C is required to maintain the permafrost. The second question is, how does permafrost form? By freezing from the surface or by burying frozen sediments? The short answer is, both. Its thickness is very variable over even short horizontal distances. It seems inconceivable that 700 m of permafrost could have come about by freezing from the surface with such an irregular result. The occurrence of ice in permafrost, such as you would expect from the burial of frozen lakes, and even frozen carcasses supports the conclusion that at least some thick permafrost is the result of burial and accumulation of frozen sediments.

If you are mathematically inclined, see Turcotte & Schubert (1982: 134–97, 332–4) for a fuller discussion of these and other points.

7 ELECTRICITY AND MAGNETISM

Electricity

We tend to think of electricity in terms of the light, heat, or music that come on at the flick of a switch. Here we are not so much concerned with these, but rather with a general understanding of the nature of electricity, and some specific applications. We are very much aware of electricity during a thunderstorm, and may become more aware of it in dry climates when static electricity is sufficient to give us a small shock when touching a door handle or putting a key in a lock.

Electricity is all around us, and in everything in our lives. Atoms consist of charged particles, negatively charged electrons and positively charged protons. Atoms remain intact because unlike charges attract and like charges repel each other. Ampère and Faraday showed that electricity and magnetism are intimately connected.

The force of attraction (unlike charges, with different signs) or repulsion (like charges, with the same signs) between two point charges in a vacuum is given by Coulomb's Law:

$$F_0 = k Q_1 Q_2 / r^2,$$

where Q_1 and Q_2 are the point charges (in units of coulomb) which may be positive or negative, r is the distance between them, and k is the proportionality constant, which equals $1/4\pi c\,\varepsilon_0 (\mathrm{N\,m^2\,C^{-2}})$. c is called the dielectric constant (dimensionless), and ε_0 is the *permittivity* of free space or of vacuum.

Electric charges are transportable and transferable, as in batteries and along wires. If you rub some materials with silk, both acquire a charge that can be taken somewhere else (with but slow dissipation). You can discharge an object by touching it, provided you are not insulated from the Earth. So there are materials that conduct electricity with ease, and others that do not. All metals are good conductors, as are some non-metals. Good conductors of electricity are also good conductors of heat, and electric insulators are also heat insulators. Most minerals are bad conductors of electricity and heat. A good conductor requires free electrons. Electricity can be made to flow down a wire and made to do work at some distance from where it is generated. An electric current flowing down a wire generates a magnetic field around it. This process is reversible, and a changing magnetic field can generate an electric current.

Some definitions

ohm The unit of resistance. It is the resistance of an electric circuit in which an electric potential difference of one volt produces a current of one amp (ampere). Or, equivalently, it is the resistance in which one watt is dissipated when a current of one amp flows through it. Its symbol is Ω.

volt Unit of potential difference and electromotive force. It is the potential difference across a resistance of one ohm when one ampere is flowing through it. Or, equivalently, it is the difference in potential between two points on a conductor carrying a current of one ampere when the power dissipated between the points is one watt.

ampere Unit of electric current (coulomb per second). It is the current produced by a potential difference of one volt across a resistance of one ohm. Or, equivalently, it is the constant current that, if maintained in two parallel, straight, conductors of infinite length and negligible circular cross-sectional area, placed 1 m apart in a vacuum, would produce a magnetic force between the two conductors equal to 20×10^6 newtons per metre of length.

watt Unit of power (joule per second). It is the power dissipated in an electric conductor carrying a current of one ampere between two points at one volt potential difference.

The main topics we are interested in are:

- the electrical conductivity/resistivity of materials, and
- the fields associated with electric currents, and the analogy with water flow.

Conductivity and resistivity; potential

Conductivity is a measure of the ease with which an electric current passes through unit volume of a substance: resistivity is the inverse. *Conductance* and *resistance* are the absolute measures of a particular volume of substance. The unit of resistance is the ohm, $\Omega(\mathrm{V\,A^{-1}})$; the unit of resistivity is the ohm-metre, ohm-m, or $\Omega\mathrm{m}(\mathrm{V\,A^{-1}\,m^2\,m^{-1}} = \mathrm{V\,A^{-1}\,m})$. So the resistance between two electrodes 2 m apart in material of resistivity $1\,\Omega\mathrm{m}$ will be $2\,\Omega$ if the cross-sectional area remains constant.

Ohm's Law states that for a given conductive material, the ratio of the voltage (potential or energy) across the ends and the current in the conductor is constant

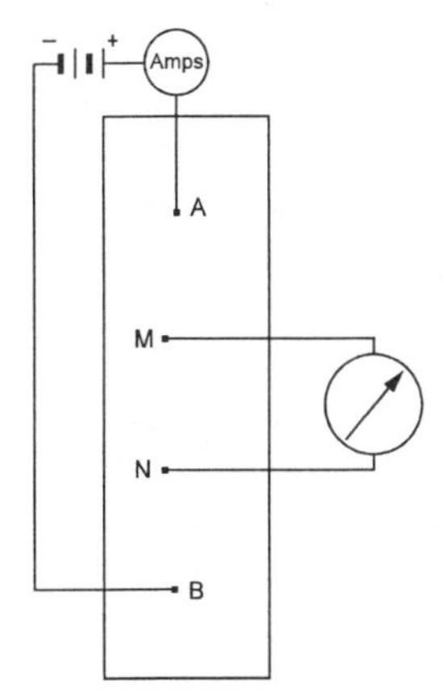

Figure 7.1 Resistivity.
The resistivity of a block of material can be measured by passing a current through two electrodes and measuring the voltage between the other two. The dimensions of the block between the electrodes are also required (Eq. 7.2).

within certain limits, and this is called the resistance of the material:

$$V\,(\text{volts})/I\,(\text{amps}) = \text{constant} = \text{resistance}, R\,(\text{ohm}). \tag{7.1}$$

If the cross-sectional area S is uniform, the resistance is proportional to the length, l, and inversely proportional to the cross-sectional area. *Resistivity* is thus related to resistance, R, by:

$$R = \rho l/S, \tag{7.2}$$

where ρ is the resistivity of the material. (These symbols are not entirely standard. R commonly stands for resistivity in electrical borehole logging studies.)

In practice, a known current is passed between two electrodes placed in the material (Fig. 7.1), and the potential difference between two others is measured. The principle of reciprocity means that if the current is passed between M and N, and the voltage (potential difference) is measured between A and B, the result will be the same. This is very important in the tools used for measuring the resistivity of rocks around boreholes.

Electric fields

When two electrodes are placed in a homogeneous, isotropic material, and a current is passed from one to the other, an electric field is created that occupies the material (Fig. 7.2). The field consists of lines of force, or flow paths, radiating out from the positive electrode A normal to the surfaces of equal potential (equipotential surfaces), and converging into the negative electrode B. There is also a line of force or a flow path from A that is diametrically opposed to the direct path to B. Indeed, the direct path is but one path, and the reason is simple. By creating a potential difference between A and B, the energy of the charges forming the elec-

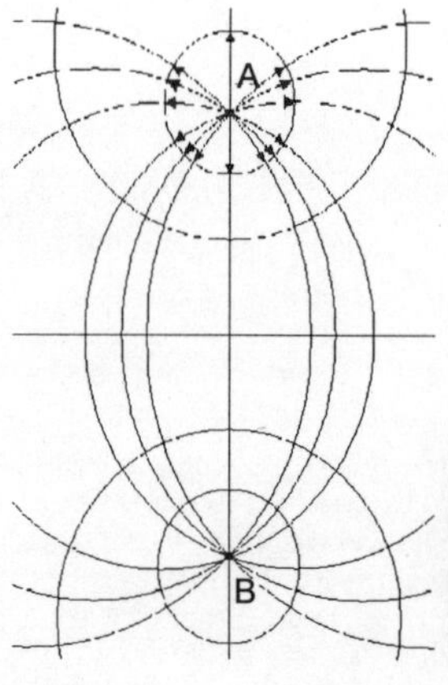

Figure 7.2 The field between two electrodes in a homogeneous isotropic medium.
Lines and surfaces of equal energy, called equipotential lines/surfaces, are normal to the flow paths (this map is part of a field, not of a block).

tric current is greatest at A, and decreases in all directions away from A. This is exactly what happens if the electrodes are replaced by water wells in a homogeneous, isotropic reservoir, pumping down at A at the same rate as pumping up at B. The energy of the water is greatest at A and water flows radially outwards, following paths to B, where the energy is least.

Electric fields exist whether or not currents exist between the electrodes and charges are moving. If currents exist (the charges are moving) there is an associated magnetic field. Consequently, if a magnetic field changes, it will cause charges to move and so generate currents.

The field can be mapped and measured if electrodes can be inserted into the medium and the potential difference measured. This is a property of the medium and of the potential difference. The same is true of groundwater. Wells drilled in an aquifer would indicate the distribution of energy or potential by virtue of the elevation to which water would rise unassisted in the boreholes (the static water level). A contour map of the elevation of the water surface in boreholes drilled to the same aquifer is a map of the energy of the aquifer. The contours are lines of equal energy (equipotential lines) and flow lines are normal to them. The surface contoured is a notional surface called the potentiometric surface.

Magnetism

Magnetism is one of the fundamental forces of the universe. The mineral magnetite, $FeO.Fe_2O_3$, commonly with titanium oxide, TiO_2, has been known for more than 2500 years for its properties of attracting iron and orientating itself in the Earth's magnetic field. A bar magnet has a *magnetic axis*, the ends of which are called *poles*. The end that seeks the geographical north when freely suspended is called the north pole or north-seeking pole (or +); the other is the south or south-seeking pole (or –).

We can magnetize bars of ferromagnetic metals (iron, nickel, cobalt, and some of their alloys) by putting them in a magnetic field; by cooling them when aligned north–south; by stroking them with another magnet (the new pole at the beginning of the stroke is the same as the stroking pole); and by hammering them when orientated in a magnetic field. This last is the reason why steel ships and boats become magnetized during their construction, particularly riveted ships[1]. All substances are magnetizable to some degree, but few are usefully magnetizable. The ferromagnetic and most other materials are paramagnetic; that is, they become magnetized with their direction of magnetization in the same direction as the magnetizing field. A few are diamagnetic (bismuth, for example) and acquire a magnetism with opposite polarity to that of the applied field.

If you cut a bar magnet in half, you get two bar magnets, each with a north and a south pole. Cut these in half and you get four magnets, and so on. It is therefore inferred that magnetized metal is magnetized on a molecular scale. Permanent magnets, as the name implies, retain their magnetism for a considerable time, but all magnets lose some of their magnetism with time. The magnetism remaining after the removal of the exciting source is called *remanent* magnetism.

Materials lose their permanent magnetic properties when heated above a temperature known as the *Curie point*. At atmospheric pressure, the Curie point of nickel is 330°C; iron, 770°C; and magnetite, 580°C. As pressure increases, so the Curie point falls.

Of the minerals, only magnetite and, to some extent, pyrrhotite are affected by a bar magnet, but many are affected by an electromagnet. The various magnetic properties of minerals allow their separation using an electromagnet.

Place a piece of thin glass on a bar magnet, and scatter iron filings on the glass. The filings orientate themselves in a pattern, showing the *field* (Figs 7.3, 7.4). Drop the magnet into iron filings, and most attach themselves to the poles of the magnet, with few near the centre.

Coulomb defined a unit magnetic pole as one that repels an identical pole placed 1 cm away in a vacuum with a force of 1 dyne (10 μN); and found that a pole of strength p repels a like pole of strength p' at a distance of r cm with a force proportional to pp'/r^2,

$$F_0 = k_0\, pp'/r^2,$$

where F_0 is in newtons, p in A m, r in m, and k_0 in N A^{-2}. The proportionality con-

1. A property that was made use of by the Germans in the Second World War, with a mine that was actuated by the ship's magnetic field. The reply was to remove the magnetism of the ship with electric cables round the ship with which the opposite field was generated – degaussing. The mines were swept by towing cables from which a strong magnetic field was generated.

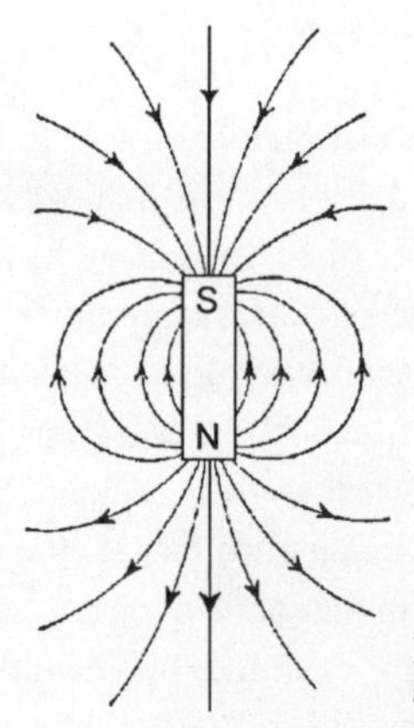

Figure 7.3 The lines of force in the field of a bar magnet.

stant $k_0 = \mu_0/4\pi = 100 \times 10^{-9} \mathrm{N A^{-2}}$ where μ_0 is the magnetic *permeability of free space* (also in units of $\mathrm{N A^{-2}}$). It will be noticed that gravity, magnetism and electricity all have similar laws, with the force of attraction proportional to $\mu_1\mu_2/r^2$, where μ is a quantity of mass, magnetic pole strength, or electric charge and r is the distance between them. The strength or intensity of a magnetic field (H) in any position is defined as the force exerted on unit north pole at that point. The unit of magnetic field strength used to be the *oersted* (oe), after the Danish physicist Hans Ørsted, defined as the strength of the field at a point where unit pole is repelled with a force of 1 dyne. The SI unit is the ampere per metre ($\mathrm{A\,m^{-1}}$), which equals $12{\cdot}566 \times 10^{-3}$ oe.

The origin of magnetism lies in electric currents, that is, moving electric charges, as first shown by Ampère. In magnetic materials, the currents are due to electrons and protons bound to the atoms. The Earth's magnetism is more complicated.

For treatment at greater depth, see Bleany (1984) and Feynman et al. (1964).

Terrestrial magnetism

The Earth has a magnetic field that is similar to that which would be generated by an axial bar magnet near its centre. This has been known for about 400 years. We can measure the Earth's magnetic field with great precision. If we suspend a magnetized needle so that it can freely orientate itself in the magnetic field, we find that it is, in general, not horizontal but has a *dip* or *inclination* below the horizontal and a *declination* or *variation* from geographic north (Fig. 7.5). The force can be resolved into three components: x, horizontal in the direction of geographic north; y, horizontal towards geographic east; and z, vertical. The force is vertical near the poles, horizontal near the Equator, but we find that the magnetic poles are not

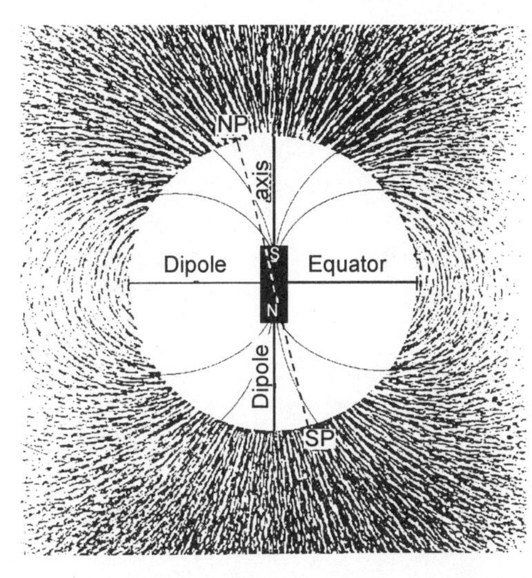

Figure 7.4 The Earth's magnetic field closely resembles that of an axial bar magnet.
NP and SP are the North Pole and the South Pole. The dipole axis does not coincide with the geographic poles. (Reproduced from *Principles of physical geology* (2nd edn) by Arthur Holmes, figure 72b, p. 988 (London: Nelson, 1988).)

at the geographic poles. The south magnetic pole in 1990 was in latitude 65·0°S, longitude 138·0°E: the north magnetic pole, in latitude 77·6°N, longitude 103·4°W – so they are not even diametrically opposite each other, but separated from the projected pole by 1370 nautical miles, or 2530 km. However, the geomagnetic poles are the poles of what is called the centred geomagnetic dipole, which is the dipole that gives the observed field most closely. In 1990, the geomagnetic poles were in 79·16°N 71·06°W, 79·16°S 108·94°E. The geomagnetic poles are opposite each other, and they were 1200 km from the geographic poles.

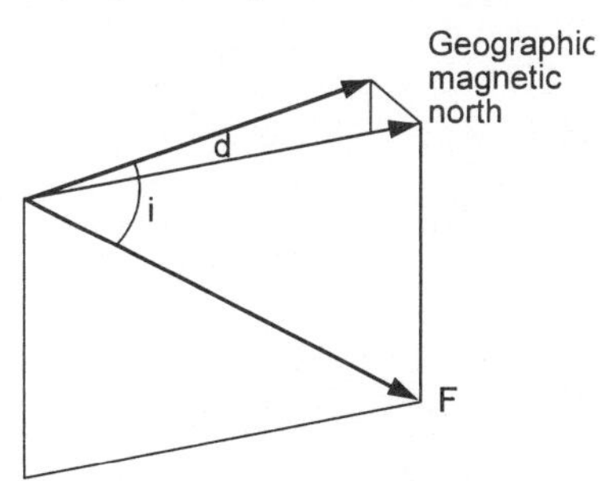

Figure 7.5 Magnetic declination or variation (*d*) and inclination or dip (*i*).

There is an approximate relationship between the magnetic dip, *i,* and the latitude relative to the *magnetic* poles:

$$\text{latitude}_{magnetic} = \tan^{-1}(\tfrac{1}{2}\tan i). \tag{7.3}$$

This is clearly only approximate because the Earth's magnetic field is neither geometrically perfect nor symmetrical. Furthermore, the indicated field, like the compass today, does not necessarily point at the north magnetic pole; it points along a field line that leads eventually to the pole. It is, however, a useful tool in palaeomagnetic studies for estimating the latitudes of continents in past ages, and polar positions.

We also find that the magnetic poles change position slowly, so the field is not constant. That is why maps and charts have a note such as "magnetic variation (1976) 13°42′ E, decreasing 11′ annually". There are also small daily, monthly, and annual changes, which are due to external influences (e.g. diurnal changes due to the Earth's rotation, and aurora australis and borealis – "Southern Lights" and "Northern Lights" – and magnetic storms).

The scientific terminology is a little different from the practical. The angle between the direction of the compass north and the north geographic pole is known as the *declination* in geophysics, *variation* in navigation and on maps. This angle varies with position, and in any one position it varies with time. The compass needle does not usually *point* to the magnetic poles but is aligned with a field line that eventually reaches the poles.

Why does the Earth have a magnetic field? Assuming that we are correct in saying that most of the interior of the Earth is above its Curie point, and that it cannot therefore be permanently magnetized, the source of the field must either lie in the shallower, low-temperature, crust or be continuously generated in the core. It is clearly something to do with the Earth's rotation for the magnetic poles to be so close to the geographic poles. It is *thought* to be caused by convection currents circulating conducting metallic fluids in the outer core, a sort of self-exciting dynamo. The sudden changes of polarity that have occurred in the past are not easily explained (sudden, that is, in the context of geological time). There is a small component, as mentioned above, caused by external influences that give rise to small changes on a short timescale, such as the magnetic storms associated with sunspot activity. These components are in the field caused by electric currents in the ionosphere and solar wind.

Remanent magnetism

Remanent magnetism is the magnetism that remains after the magnetizing force has been removed or changed. In rocks there are minerals that have magnetic properties, such as magnetite (naturally!), that may have been magnetized in the past by a terrestrial magnetic field with its poles in a different position or reversed. Today, such an object is in the present field, the strength and direction of which is accurately known. Using a technique for removing the magnetization due to the present terrestrial field, the remanent magnetism can be determined.

When a ferrimagnetic material cools from above to below its Curie point, as some lavas do, it acquires its magnetism from the local field. This is called *ther-*

moremanent magnetism. If new magnetic minerals grow during diagenesis or metamorphism, they become permanently magnetized to a measurable degree once they attain a certain size. If magnetized sedimentary particles settle from suspension in still water, or are agitated gently on the bottom before accumulating into the stratigraphic record, they may orientate themselves in the geomagnetic field and so preserve a record of it. All these are examples of remanent magnetism. Remanent magnetism is the basis of *palaeomagnetism*, the elucidation of past magnetic fields and of sea-floor spreading, continental drift and plate tectonics.

8
STRESS AND STRAIN

Pressure, p, is a surface force divided by the area on which the force is exerted $[MLT^{-2}/L^2 = ML^{-1}T^{-2}]$. Stress, S or σ, is also a force divided by the area it acts on, but it is three dimensional. More properly in some contexts, pressure and stress are the limiting value of the ratio as the area becomes very small,

$$S = \lim_{\delta A \to \infty} (\delta F / \delta A) \ . \tag{8.1}$$

Force, as we saw on page 13, is a vector and we can resolve a force into its x, y, z components by vector arithmetic. A surface force can be resolved into a normal component on the surface, and a shear component in the surface. Can we resolve a stress? Yes, but not by vector arithmetic. In fact, six components are needed to define a stress completely, and stresses are *tensors. Tensors cannot be resolved by vector arithmetic.*

Take a cube and apply a pressure between two opposing faces. Within the cube there will be a state of stress in three dimensions. More generally, the cube may have any orientation with respect to the force applied. On each face there will be a normal and two possible shear directions in the coordinate system, giving 18 in all. The cube is in equilibrium, so we need only consider the forces acting on three sides, giving nine components as in the following matrix:

$\sigma_x \quad \tau_{xy} \quad \tau_{xz}$ acting along the x axis

$\tau_{yx} \quad \sigma_y \quad \tau_{yz}$ acting along the y axis

$\tau_{zx} \quad \tau_{zy} \quad \sigma_z$ acting along the z axis

These further reduce to six quantities because $\tau_{xy} = \tau_{yx}$, $\tau_{xz} = \tau_{zx}$, and $\tau_{yz} = \tau_{zy}$ when the object is not subjected to a net turning couple.

The *stress ellipsoid* is a convenient way of describing the state of stress in a solid. The stress ellipsoid is defined by three mutually perpendicular axes and the stresses along them. The axes are called the principal directions, and the magnitude of the stresses in these directions are called the principal stresses, designated

σ_x, σ_y and σ_z (*z* being, by convention, vertical). These three principal stresses may be of any magnitude, but they are by definition without shear components in the plane normal to the principal directions. The greatest principal stress is given the symbol σ_1; the least, σ_3; and the intermediate, σ_2. In geology the principal stresses are usually compressive, so if a principal stress is not called a principal compressive stress, compressive is usually understood.

Stresses are compressive when the strain tends to reduce dimensions in the direction of the stress, tensional when the strain tends to increase the dimensions. Great care is needed with the concept of tension in geology because real tension is rare. If a large volume of rock underlying flat topography has no force applied to it other than that of gravity, the principal stresses will be vertical and horizontal, approximately equal in magnitude or intensity, and all compressive below a depth of perhaps a few tens of metres. The state in which $\sigma_x = \sigma_y = \sigma_z$ is called *hydrostatic*. This word, relating to standing water, should have been left to fluid mechanics, but it was not and so we must be careful to avoid ambiguity in its use.

If another stress is imposed in addition to gravity on this same volume of rock, the principal compressive stresses will no longer be equal and we shall designate them σ_1, σ_2, and σ_3. We can also describe the principal stresses in terms of their deviation from the mean stress, $(\sigma_1 + \sigma_2 + \sigma_3)/3$. These differences between the actual and the mean principal stresses are then called the *deviatoric stresses*. If the stress field is not hydrostatic, at least one deviatoric stress will always be tensional, but the real stress will only very rarely be tensional. What geologists loosely call "tensional faults" are almost always formed in a stress field that is entirely compressional, but with a sufficiently large difference existing (or having existed) between the greatest and the least compressive stresses. Such a field produces extensional strain relative to the mean stress state, which can be regarded as hydrostatic.

The reason why tensors cannot be resolved by vector arithmetic will be apparent. Consider a force acting along the *z* axis, in the *z* principal direction. The component of *force* parallel to the *x* or *y* axis is zero by definition, but the *stresses* along those axes may be of any magnitude.

Strain, as we noted above, is the change of shape or volume of a body as a result of stress. The application of forces may also result in a change of position, or translation, of the body as well as strain in the body. The relationship between stress and strain involves the physical properties of the material stressed.

Elasticity and the elastic moduli

Elasticity is the property a material may have of recovering its original shape and size "immediately" after a deforming force has been removed. Not all materials have the same degree of elasticity. Rubber, for example, is very elastic while lead is not. Elasticity arises from the forces between atoms or molecules and shows that the forces are repulsive when the atoms are closer than some natural spacing, and attractive when they are farther away. In the context of geology, time comes into this and relatively few materials would be elastic if the deforming force or load were to be removed only after several millions of years. We see this in folded rocks and compacted sediments, for example. The elastic moduli quantify the elastic properties of materials. They are material constants relating the amount of the physical deformation on a material to the amount of force producing it. The moduli are important for several reasons. Rocks have characteristic values of the moduli, so the moduli help in the analysis of deformation of rocks under stress. This is particularly important in the derivation of the wave equations that govern the passage of elastic waves through the rocks, as in earthquakes and seismic surveys. The speed of propagation of elastic waves is also a function of an elastic modulus. Small displacements on a short timescale can be regarded as perfectly elastic.

Robert Hooke found in 1678 that the extension of a spring was proportional to the force applied to it – that is, strain (ε) was proportional to stress (σ) in the spring (strain being defined as the extension per unit of length, $\delta l/l$). This is Hooke's Law. The ratio σ/ε is a material constant and is the basis for some weighing devices. Hooke's Law only applies to small weights and a short time. Larger weights and longer time result in failure of the weighing device to return to zero, or failure of the stressed material to return to its original unstressed length. It is then said that the *proportional limit* or *elastic limit* of the material has been exceeded. These two terms are not strictly synonymous: the proportional limit simply marks the limit of Hooke's law – there will be some elastic recovery, but not complete –while the elastic limit marks just that, the limit of elastic behaviour and the beginning of plastic deformation.

The one-dimensional relationship of Hooke's law can be extended to three dimensions with the assumption that each component of stress is related linearly to the components of strain, and that we have a perfectly isotropic elastic solid. Taking principal stresses and strains, this assumption leads to

$$\sigma_1 = (\lambda+2G)\varepsilon_1+\lambda\varepsilon_2+\lambda\varepsilon_3$$

$$\sigma_2 = \lambda\varepsilon_1+(\lambda+2G)\varepsilon_2+\lambda\varepsilon_3$$

$$\sigma_3 = \lambda\varepsilon_1+\lambda\varepsilon_2+(\lambda+2G)\varepsilon_3, \qquad (8.2a)$$

where λ and G are known as Lamé's parameters or Lamé's constants (after the French mathematician and engineer Gabriel Lamé, 1795–1870). The constant – it is a material constant – or parameter $(\lambda+2G)$ relates strain to stress in one direction, while λ relates strain to stress in the two perpendicular directions. G and λ have the dimensions of pressure and units of pascals. Lamé's parameters are important also because the speed or velocity of elastic waves in rocks is dependent on them, as we shall see below.

The volumetric strain, $\varepsilon_1+\varepsilon_2+\varepsilon_3$, is called the dilatation (*sic!*) and is represented by Δ usually in geological and rock-mechanical work, so the Equations 8.2a above can be written

$$\sigma_1 = \lambda\Delta+2G\varepsilon_1$$
$$\sigma_2 = \lambda\Delta+2G\varepsilon_2$$
$$\sigma_3 = \lambda\Delta+2G\varepsilon_3. \quad (8.2b)$$

The coordinate system may not be parallel to the principal stresses and strains. It can be shown that

$$\sigma_x = \lambda\Delta+2G\varepsilon_x$$
$$\sigma_y = \lambda\Delta+2G\varepsilon_y$$
$$\sigma_z = \lambda\Delta+2G\varepsilon_z. \quad (8.3a)$$

Since these are not necessarily principal stresses, there will be three possible shear stresses

$$\tau_{yz} = G\gamma_{yz} \quad \tau_{zx} = G\gamma_{zx} \quad \tau_{xy} = G\gamma_{xy} \quad (8.3b)$$

where γ is the shear strain. Adding Equations 8.3a and using the dilatation term above,

$$\sigma_x+\sigma_y+\sigma_z = (3\lambda+2G)\Delta. \quad (8.4)$$

There are other material constants of a similar nature. Consider a wire that is not coiled supporting a weight mg that is not so large that the proportional or elastic limit will be exceeded. The ratio of stress to strain is constant, so $mg/A = E/(\delta l/l)$, where A is the cross-sectional area of the wire. E is called *Young's modulus*. It is a special case of Hooke's law and is sometimes called the modulus of elasticity. The dimensions of Young's modulus are those of a force on an area (a pressure), $ML^{-1}T^{-2}$, and the unit is the pascal (N m^{-2}). It is essentially uniaxial.

The ratio of shear stress to shear strain, τ/γ, is called the *modulus of rigidity* or *shear modulus* (G). Strain is dimensionless, being $\delta l/l$, so the dimensions are also those of a pressure, and the unit is the pascal.

The wire also suffers a loss of radius when extended. The modulus that relates the loss of radius to the increase of length, is called *Poisson's ratio*: $v = (\delta r/r)/(\delta l/l)$. This is dimensionless.

When an elastic body is submerged and a pressure is applied to the fluid enveloping the body, there is a change of volume. The strain is defined as a change of volume per unit of volume, so the ratio of the change of pressure and the strain it causes is $\delta p/(\delta V/V)$. This is the *bulk modulus* (K), and it has the dimensions of pressure. *Compressibility* is the inverse of K.

These three elastic constants are related by

$$E = 3K(1-2v) = 2G(1+v)$$

and each can be expressed in terms of Lamé's parameters:

Young's modulus of elasticity: $E = \sigma_1/\varepsilon_1 = G(3\lambda+2G)/(\lambda+G)$ $[ML^{-1}T^{-2}]$

Modulus of rigidity, or *shear modulus:* $G = \tau/\gamma$ $[ML^{-1}T^{-1}]$

Poisson's ratio: $v = \varepsilon_2/\varepsilon_1 = \gamma/2(\gamma+G)$ [dimensionless] or [0]

Bulk modulus: $K = \sigma/\Delta = (1+2G/3)$. $[ML^{-1}T^{-2}]$

See Jaeger & Cook (1979) for a full analysis of these constants and their derivation, and the propagation of elastic waves through rocks; and Ramsay (1967: 283ff.).

Friction

The laws of thermodynamics were once wittily paraphrased as

You cannot win: at best you can draw.

You can only draw if you can get to absolute zero (0 K)

You cannot get to absolute zero.

Friction is a force that dissipates all forms of kinetic energy, transforming some of it to heat. Even in space (but not in a perfect vacuum) artificial satellites lose their kinetic energy very slowly, and eventually will return to Earth. The frictional resistance on return to the atmosphere slows the satellite down and generates spectacular heat against which passengers and sensitive equipment must be shielded. The total energy is conserved.

Friction is involved in the flow of fluids, gases to a lesser extent than liquids. This is the practical meaning of viscosity. Friction is critical in the design of aeroplanes because that is the main loss of energy in flight. It is also important in the design of motor cars because that is the main loss of energy as one increases speed. In both cases, the changing fuel consumption with speed, per unit of distance travelled, is a measure of the increasing losses due to frictional resistance of the air. Curiously, air resistance, whatever the shape of the solid, generally increases as the square of the speed or velocity.

Friction is involved in sliding, which we shall consider in some detail a little later. You can place a book on a bench or any other planar surface and lift one end of the bench. There is some angle of slope of the bench at which the book will just begin to slide, so the frictional force can be measured. It is equal to the maximum component of weight down the slope just before the object begins to slide, W $\sin\theta$ (where θ is the angle of slope below the horizontal). This is sometimes called the limiting or static friction.

The coefficient of friction, μ, is the ratio of the limiting friction to the component of weight normal to the sliding surface,

$$\mu = W\sin\theta / W\cos\theta = \tan\theta, \tag{8.5}$$

and is independent of the size or weight of the object. For solids, this is determined for dry materials without any form of lubricant at the sliding surface. For porous materials it can be shown that the coefficient of friction is the same for dry materials in air as wet materials in water (the requirement of non-lubricating fluids remains). It is the *ambient* fluid that matters, as long as it is not a lubricant.

The coefficient of friction is a material constant, for practical purposes. Once an object is sliding, it requires less force to keep it sliding, so the coefficient of friction is smaller. This is called the *kinetic* coefficient of friction.

Sliding down a slope is not the only form of sliding: you can push a block along a surface. If you push a block along a horizontal surface, you find that the force required to start movement has a fairly constant relationship to the normal reaction (i.e. the normal force between the two surfaces, W),

$$F = \mu W. \tag{8.6}$$

Note: if dense materials are given highly polished surfaces, sliding does not readily take place because of the molecular attraction between the surfaces. Two highly polished flat surfaces of metal are not easily separated, and you may be able to lift both when you lift the upper one.

When we strain a solid to the point of fracture, one surface slides over another within the material. Clearly there is a *cohesive strength* that must be overcome before fracture occurs; then there is frictional resistance to be overcome. The angle of fracture is a function of both of these components. This topic will be taken up later (see **fracture** on p. 87).

Viscosity

Viscosity is the property of internal friction in fluids. There are two kinds of viscosity, so care must be taken to distinguish between them. We can also talk of the *effective viscosity* of rocks because, in the immense timescale of geology, rocks

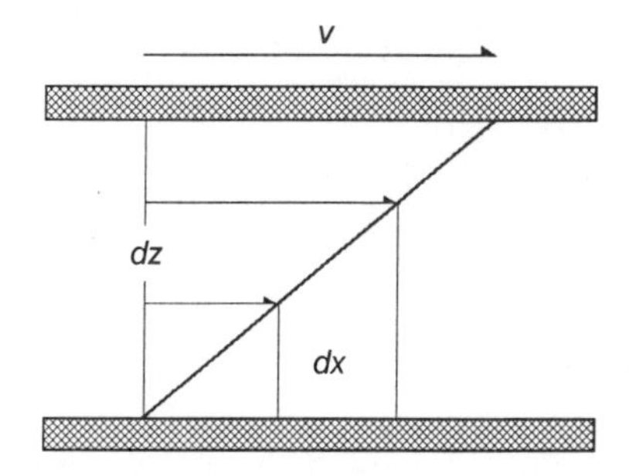

Figure 8.1 Newton's concept of viscosity (in section).
When the upper plate moves relative to the lower, the lubricant between them is assumed to have adhered to each plate and to have sheared linearly between them.

can behave much like fluids – but it would be unwise to assume that the internal motion of rocks and fluids are strictly comparable.

Coefficient of viscosity; absolute or dynamic viscosity Consider a liquid between two closely spaced parallel plates (Fig. 8.1) of which the top plate is moving with constant velocity, V, relative to the lower plate. Fluid adheres to each plate. The velocity of the fluid against the upper plate is V, while the fluid adhering to the lower plate does not move. Between the two, it is reasonable to assume that successive layers slip relative to those below and above it, so that the velocity is linearly distributed between the two plates. This is Newton's concept of viscosity.

The *coefficient of viscosity*, η, is defined as the ratio of shear stress to the rate of shear strain:

$$\eta = \tau/(dV/dz). \qquad (8.7)$$

In words, the coefficient of viscosity is the tangential force per unit area that maintains unit relative velocity between two parallel planes unit distance apart.

The coefficient of viscosity is also known as *absolute* or *dynamic* viscosity. The SI unit is *newton second per square metre* ($N\,s\,m^{-2}$). Other units are the *poise*, P, which is grams per centimetre second ($g\,cm^{-1}\,s^{-1}$) in units of mass, length and time (dyne second per square centimetre in units of force, length and time). The poise is named after the French physician, J. L. M. Poiseuille (1799–1869) whose interest in blood flow led to an elegant series of experiments on flow through small pipes.

If the viscosity remains constant throughout the fluid, and at different strain rates, the fluid is said to be *newtonian*. Non-newtonian fluids are those with variable viscosity.

Kinematic viscosity We find in many expressions in fluid mechanics the ratio of the dynamic viscosity of a fluid to its mass density, η/ρ. It has the dimensions L^2T^{-1}, with the symbol ν (the Greek letter nu). It is known as kinematic viscosity because it concerns motion without reference to force. The unit is $m^2\,s^{-1}$ in SI, or the *stoke*, which is $cm^2\,s^{-1}$. The stoke is named after Sir George Stokes

(1819–1903), the British mathematician and physicist, whose work on the internal friction of fluids and the motion of pendulums led to what is now called Stokes' Law for the terminal velocity of a small sphere falling through a fluid.

Sliding

Sliding takes place in several geological processes. When rocks are faulted, sliding takes place in the fault plane. When a layered sequence of rocks is folded, some adjustment of the beds is likely to take place by sliding along bedding surfaces. When mountain ranges are created, slopes may be generated that are steep enough for rock sequences to slide down the slope. This is not just a small-scale phenomenon, or a new conception. Nearly 100 years ago, Törnebohm postulated movement of blocks at least 130 km long on Caledonian thrusts in Scandinavia. Sliding on such a scale raised an interesting question: the strength of the rocks themselves limited to a few kilometres the length of block that could be pushed, but if sliding down a slope is required, then the 65 km difference of elevation of a block 130 km long down a slope of 30°, which seemed to be required, was unacceptable – and there was no geological evidence for such a slope. The resolution of this paradox lies in the nature of sliding.

There are two sorts of sliding: lubricated and unlubricated. We shall take unlubricated sliding first, for better understanding of the two.

Unlubricated sliding

When a rectangular block of thickness h is placed on a plane surface that is not sloping steeply enough for the block to slide, its weight σ_z in the ambient fluid, per unit area of the base, is

$$\sigma_z = \rho_b gh - \rho_a gh = (\rho_b - \rho_a)\, gh, \tag{8.8a}$$

where the suffix a refers to the ambient fluid. This can be resolved into a component normal to the surface (Fig. 8.2)

$$\sigma_n = (\rho_b - \rho_a) gh\cos\theta \tag{8.8b}$$

and a shear component parallel to the surface

$$T = (\rho_b - \rho_a) gh\sin\theta. \tag{8.8c}$$

These may be regarded as active stresses. They give rise to reactive stresses, a normal reaction, and frictional resistance (τ) that tends to prevent sliding. As the

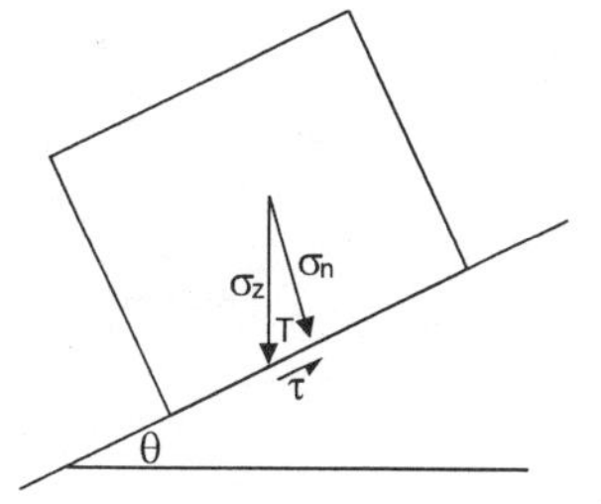

Figure 8.2 Components of the weight of a block. The vertical component is σ_z, that normal to the sliding surface, σ_n, and that parallel to the sliding surface, *T*. The frictional resistance to sliding is τ.

angle of slope θ increases, so the shear component of weight increases. There is some critical value of θ at which the shear component of weight is equal to the frictional resistance and the block is just about to slide. The friction is due to adhesion or cohesion between bodies, or to the roughness of the surfaces that would be sliding surfaces. The roughness can only be overcome by deformation or fracture of the protuberances.

The Coulomb or Mohr–Coulomb criterion for simple unlubricated sliding (see Hubbert 1951: 363) is

$$\tau = \tau_0 + \sigma_n \tan\phi \qquad (8.9)$$

where τ_0 is the *cohesive strength* or *initial shear strength* of the material at the surface that will become the sliding surface, when the normal stress σ_n is zero; and $\tan\phi$ is the coefficient of sliding friction. Sliding will therefore take place when

$$T = \tau,$$

$$(\rho_b - \rho_a) gh \sin\theta = \tau_o + (\rho_b - \rho_a)\, gh \cos\theta \tan\phi$$

(from Equations 8.8b, 8.8c, 8.9) and

$$\tan\theta = (\tau_o/(\rho_b - \rho_a) gh \cos\theta) + \tan\phi. \qquad (8.10)$$

This equation shows that unlubricated sliding will usually take place on a slope θ that is rather greater than ϕ.

The effect of pore-fluid pressure can be very important, reducing the effective stress and greatly reducing the angle at which sliding can take place. This was studied extensively by Terzaghi (1943: 235; 1950) and Hubbert & Rubey (1959). To pursue this topic further, it may help to read Chapman (1979) first.

Lubricated sliding

Following Kehle (1970), suppose an extensive, relatively thin sheet of a single liquid is flowing in uniform laminar flow down a gentle slope on a planar surface (Fig. 8.3) and can be regarded as being newtonian (see p. 3), then

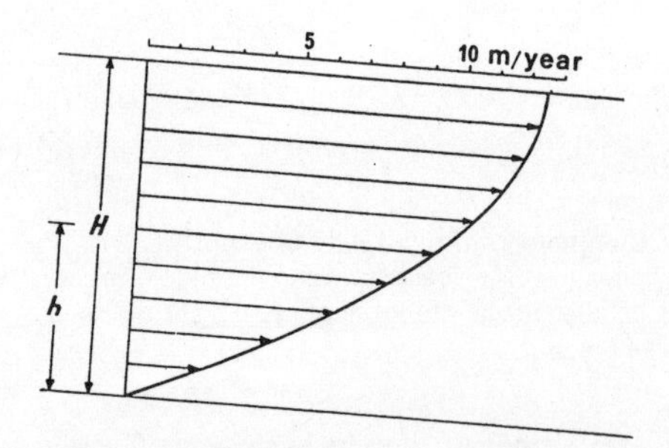

Figure 8.3 Idealized velocity profile through 2 km mudrock on a slope of 5°.
The assumed properties of the mudrock are: mass density 2000 kg m^{-3}, viscosity 10×10^{15} Pas.

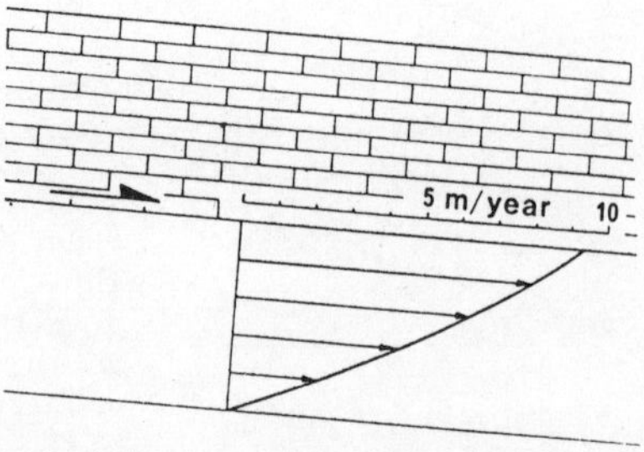

Figure 8.4 Lubricated sliding (idealized).
The mudrock has the same properties as Figure 8.3.

$$\tau = \eta \, dV/dh. \tag{8.11}$$

Assume that

$$T = \tau = (\rho_b - \rho_a)\, g\, (H-h)\sin\theta \tag{8.12}$$

where H is the total thickness of liquid flowing and $H-h$ is the thickness contributing to the shear stress at the level of interest. Note that the shear stress is maximum at the base, where the velocity is zero; and zero at the top where the velocity is greatest. Equating Equations 8.11 and 8.12,

$$\eta \, dV/dh = (\rho_b - \rho_a)\, g\, (H-h)\sin\theta \tag{8.13}$$

and integrating with respect to h (noting that $V = 0$ when $h = 0$),

$$V = [(\rho_b - \rho_a)/\eta]\, g\, (Hh - h^2/2)\sin\theta. \tag{8.14}$$

If we now consider half of the thickness H to be composed of rigid material of the same density (Fig. 8.4), its sliding velocity will be the same as at half the thickness of original lubricant.

In geology, viscosities will be very large, probably of the order of 10×10^{15} Pa s, but even these could lead to catastrophic rates of sliding of perhaps 10 m a year on quite small slopes. When Smoluchowski (1909) addressed the problems of sliding he wrote:

> Suppose a layer of plastic material, say pitch, interposed between the block and the underlying bed; or suppose the bed to be composed of such material: then the law of viscous liquid friction will come into play, instead of the friction of solids; therefore any force, however small, will succeed in moving the block. Its velocity may be small if the plasticity is small, but in geology we have plenty of time; there is no hurry.

Bending and folding

There are several ways of bending layered material, and it is one of the retarding influences on geology that many seem to think of bending only in terms of lateral compression (Fig. 8.5). Carey's early papers in this regard are still worth reading (Carey 1954). (Since we are concerned with geology, we shall talk about *folding* or *deformation* rather than bending.)

Folding is certainly possible by crumpling a thin sheet under lateral compression, but it arises more commonly from mechanical instability in the outer parts of the Earth, and in layered sequences under conditions where gravity plays a part. In an unstable layered system (such as one in which the upper layers are more dense than the lower), deformation will result in folding of the type shown in the upper part of Figure 8.5 – and such folding may be induced in stable layers above and below the unstable.

Folding requires compensation in space, either by sliding on surfaces, or flowing of the less viscous material, or both. Bending a beam puts the concave side in compression and the convex side in extension. Under some circumstances this can cause fracture, but in some cases natural rocks cannot sustain such stresses over long periods of time, and behave more like liquids.

Fracture

We follow Otto Mohr (1882), and Hubbert (1961). Consider a rectangular solid in which the principal stresses are approximately equal. If a compressive force is applied between two parallel faces, the principal stress normal to those faces will increase; there will then be a maximum principal stress, a minimum principal stress and an intermediate principal stress, which we shall denote by σ_1, σ_3, and σ_2 respectively.

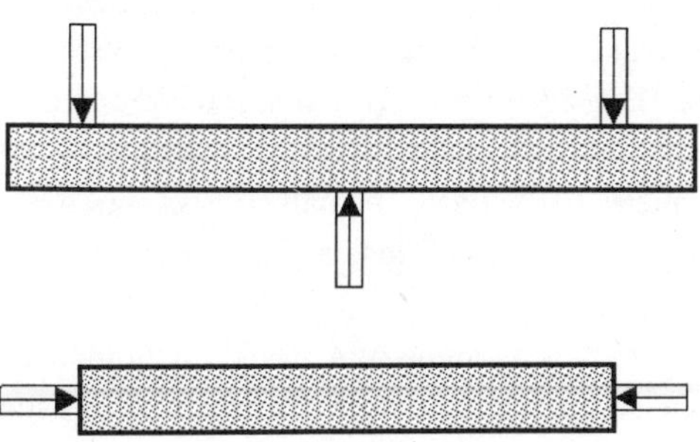

Figure 8.5 Folding can be by bending (top) or buckling (foot).

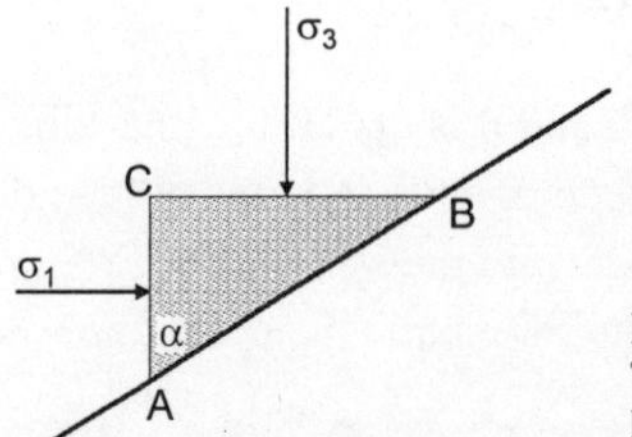

Figure 8.6 Stresses on a small prism of rock. The stresses are across a surface that lies at an angle α to the least principal stress, σ_3.

Consider a small prism of unit width on an arbitrary plane (Fig. 8.6), the prism being small enough for its weight to be negligible in the whole, but not so small that it is no longer mechanically representative of the whole. We shall assume that only σ_1 and σ_3 are significant in the analysis, σ_2 being parallel to the arbitrary plane. Let σ_n be the normal stress on the surface, and τ the shear component along the surface. Let the side AB have unit length, then the sum of the vertical and horizontal components of the forces acting on the prism are, respectively,

$$\sigma_1 \cos\alpha - \sigma_n \cos\alpha - \tau \sin\alpha = 0;$$

$$\sigma_3 \sin\alpha - \sigma_n \sin\alpha + t \cos\alpha = 0.$$

Solving these for σ_n and τ,

$$\sigma_n = \sigma_1 \cos^2 \alpha + \sigma_3 \sin^2 \alpha;$$

$$\tau = (\sigma_1 - \sigma_3) \sin\alpha \cos\alpha. \qquad (8.15a)$$

These can be reduced to the more useful form

$$\sigma_n = ½(\sigma_1 + \sigma_3) + ½(\sigma_1 - \sigma_3) \cos 2\alpha;$$

$$\tau = ½(\sigma_1 - \sigma_3) \sin 2\alpha. \qquad (8.15b)$$

Mohr represented these relationships in a simple geometrical construction that has become known as *Mohr's circle* (Fig. 8.7). From this we see that the shear stress is maximum when $2\alpha = \pm 90°$ and $\alpha = \pm 45°$, and $\tau_{max} = ½(\sigma_1 - \sigma_3)$. But the material may fail with the shear stress less than the maximum. We return to the Coulomb criterion for sliding (Eqn 8.9), $\tau = \tau_0 + \sigma_n \tan\phi$. If the properties of the material are known, this equation can be plotted on the Mohr diagram. (The Coulomb criterion is not necessarily linear, particularly at small stresses, but it is a good approximation for many materials.) The conditions at fracture are given when the Coulomb criterion is tangential to Mohr's circle. The slope of the criterion is ϕ and –ϕ, so the plane of fracture makes an angle of 45° ±ϕ/2 with the axis of least principal stress, σ_3. It is because the internal angle of friction, ϕ, is usually close to 30° in rocks, and because the principal directions tend to be vertical and horizontal, that faults are commonly inclined at about 30° or 60° to the vertical for normal and thrust faults respectively.

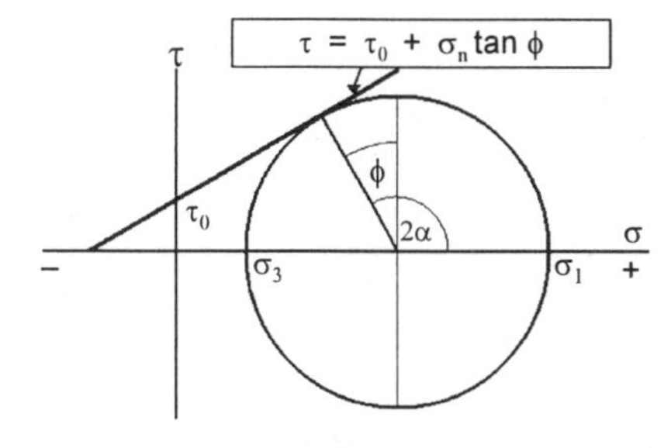

Figure 8.7 Mohr's circle.
The stress to the left of the τ-axis is tensional, that to the right, compressional. The diameter of the circle is the difference between the greatest and the least principal stresses. The line tangential to this circle is Coulomb's criterion, the slope of which is ϕ.This angle is also the angle of internal friction of the material.

Compaction, consolidation, lithification

Compaction is the reduction of the bulk volume of a sediment or rock by the reduction of pore space in it. *Consolidation* is any process that changes loose sediment to a coherent rock (or magma or lava to a solid rock, but we are more concerned with sediment here). It embraces compaction of sediment but includes the addition of solid substances to what was pore space – cementation. Compaction is essentially gravitational; consolidation includes the deposition of cements. *Lithification* is a synonym for consolidation.

It is reasonable to suppose that compaction, the loss of porosity due to compression of a particulate medium, is a function of porosity (f) or bulk density, and that the rate of loss of porosity with depth (z) is proportional to porosity; that is,

$$df/dz = -cf = -f/b,$$

from which we obtain by integration (remembering that the boundary conditions are that the sedimentary rock had an initial porosity, and that the porosity cannot be reduced below zero),

$$f = f_0 \, e^{-z/b}. \qquad (8.16)$$

The quantity b is called the *scale length* [L] (it is more convenient to divide by a large number than multiply by a small one), and f_0 is the porosity at zero depth (which can be interpreted as the depth at which the sediment accumulated into the stratigraphic record). The value of b can be obtained from the data by solving Equation 8.16 for b. Equally, it is the depth at which the porosity of normally compacted mudrock reaches the value f_0/e (or $0{\cdot}368\, f_0$). The value of f_0 is usually about 0·5, so b is usually about the depth at which about 18 per cent porosity is achieved in normal compaction.

This equation is found to be a good summary of the compaction of mudrocks, with scale lengths varying from about 500m to 4km. When the scale length is large, the curve approaches a straight line, and it could be that sandstone compaction, which appears to be linear with depth, follows the same laws with a very large scale length.

Bulk density (ρ_b) is related to porosity by

$$\rho_b = f\rho_f + (1-f)\,\rho_s \tag{8.17}$$

where ρ_f and ρ_s refer to the mean mass densities of the pore fluid and of the solids.

Let us digress briefly to a practical application. In boreholes, it is much easier to measure electric or acoustic properties of rocks than it is to measure their porosity, but we are often more interested in the porosity. We are therefore concerned with the search for formulae relating one property with another. What then is the relationship between mudrock porosity and the sonic velocity, or its inverse, called the transit time (symbol Δt_{sh}) [$L^{-1}T$, inverse velocity] in mudrock?

The boundary conditions are that there will be some maximum value of porosity, f_0, when the mudrock first accumulates into the stratigraphic record. This is usually close to 50 per cent. (Note that we are not concerned with the superficial porosities of 70 per cent and more in mud that has yet to accumulate into the stratigraphic record.) And the minimum porosity at depth will be close to zero. There will be a corresponding transit time, Δt_0, near the surface, and a limiting value of the transit time corresponding to zero porosity. This latter is called the matrix transit time, Δt_{matrix} or Δt_{ma}.

When seeking a relationship between two quantities of different dimensions, in this case L^3/L^3 and $L^{-1}T$, it may be assumed that a dimensionless form will be required, and that the dimensionless porosity, f/f_0, will be related to a dimensionless transit time involving Δt_{sh}, Δt_0, and Δt_{ma}. The boundary conditions are that when $f = f_0$, $\Delta t_{sh} = \Delta t_0$; and when $f = 0$, $\Delta t_{sh} = \Delta t_{ma}$. Assuming simple linearity, the following expression satisfies these conditions:

$$f/f_0 = (\Delta t_{sh} - \Delta t_{ma})/(\Delta t_0 - \Delta t_{ma}). \tag{8.18}$$

If you now plot Δt_{sh} against depth, a curve results. If you plot log Δt_{sh} against depth, an approximately linear trend results, suggesting an expression in the form

$$\Delta t_{sh} = \Delta t_0 e^{-z/b},$$

analogous to that derived for the compaction of mudrock above (see Eq. 8.16).

Although this has been used for years, and gives satisfactory results to depths of 2 or 3 km, it *must* be wrong because it does not satisfy the boundary conditions. As the depth becomes very large relative to the scale length b, $\Delta t_{sh} \to 0$, whereas it should tend towards Δt_{ma}. In other words, the function should give a curve that is asymptotic to Δt_{ma} at depth.

Substituting Equation 8.16 into Equation 8.18, we obtain

$$(\Delta t_{sh} - \Delta t_{ma})/(\Delta t_0 - \Delta t_{ma}) = f/f_0 = e^{-z/b},$$

which leads to

$$\Delta t_{sh} = (\Delta t_0 - \Delta t_{ma})\, e^{-z/b} + \Delta t_{ma}. \qquad (8.19)$$

Observed values of Δt_0 in shales or mudrocks are 555 μs m^{-1} (170 μs ft^{-1}), and of Δt_{ma}, 180 μs m^{-1} (55 μs ft^{-1}); and the boundary conditions are satisfied.

9 SEA WAVES

Waves on water are not the same sort of waves as sound or light – mainly because water is relatively incompressible. There are two main sorts of water waves: larger waves that are gravitational, and ripples that are due more to surface tension.

When the wind blows over deep, smooth, water, friction tends to push the surface in the same direction, and irregularities give rise to waves that move in the same direction as the wind, with their crests normal to it. The first clear waves are of short wavelength, small height, but relatively steep. The ratio of wave height to wave length (H/λ), called the steepness, is about 1:12 for these new waves. As the wind continues to blow, the waves become higher and longer (initially retaining their steepness); but as the wavelength increases, so does the speed. This acceleration reduces the relative wind strength (velocity) so the rate of increase of height decreases, and the rate of increase of wavelength decreases. They become less steep (about 1:25), but the *period* (the time between the passage of successive crests through one position, the inverse of frequency) remains constant.

The maximum height (H_{max}) and the maximum wavelength (λ_{max}) generated by strong wind blowing in one direction at W knots or $m\,s^{-1}$ for a couple of days over a fetch of several hundred kilometres are given approximately by the following formulae:

$$H_{max} = W^2/165 \text{ (m, knots)} \qquad \lambda_{max} = W^2/6{\cdot}6 \text{ (m, knots)}$$

$$H_{max} = W^2/45 \text{ (m, m s}^{-1}\text{)} \qquad \lambda_{max} = W^2/1{\cdot}74 \text{ (m, m s}^{-1}\text{)}.$$

What is the relationship between wavelength and speed? It could be a function of viscosity (probably the dimensionless ratio of viscosities of air and water), g, λ, ρ (the water mass density), and the function must have the dimensions of velocity, LT^{-1}. Using elementary dimensional analysis,

$$c = f(g, \lambda, \rho)$$

and

$$LT^{-1} = (LT^{-2})^a (L)^b (ML^{-3})^c.$$

Solving for M: $\quad 0 = c \quad$ (so density is not a component)

For L: $\quad 1 = a + b$

For T: $\quad -1 = -2a,$

from which $a = b = \frac{1}{2}$, and the function is of the form

$$c = \text{constant } (g\lambda)^{1/2}.$$

For gravity waves, it turns out that the constant equals $(2\pi)^{-1/2}$, so

$$c = (g\lambda/2\pi)^{1/2} = 1{\cdot}25\ \sqrt{\lambda}\,\mathrm{m\,s^{-1}}. \tag{9.1}$$

There is a relationship between period (Ts) and the velocity ($c\,\mathrm{m\,s^{-1}}$) and the wavelength (λm):

$$T = \lambda/c. \tag{9.2}$$

So, $c = 1{\cdot}56\ T\ \ \mathrm{m\,s^{-1}}$

$c = 1{\cdot}25\ \sqrt{\lambda}\ \mathrm{m\,s^{-1}}$

$1 = 1{\cdot}56\ T^2\ \ \mathrm{m}.$

A yacht hove-to in a storm can measure T quite accurately (it is more difficult if moving at an irregular speed), and so λ can be estimated. A period of 10s, for example, implies a wavelength of 156m and a velocity of 15½$\mathrm{m\,s^{-1}}$ (56$\mathrm{km\,h^{-1}}$, 30 knots).

Waves normally move in a group, each wave travelling at $(g\lambda/2\pi)^{1/2}\mathrm{m\,s^{-1}}$. The waves at the front of a group tend to die out because their energy is dissipated, and new ones form at the back. The group as a whole travels more slowly than the individual waves. The *group velocity* is important because the energy of the waves depends on the group velocity, not the wave velocity. The group velocity of gravity waves is half the wave velocity (see Note 3).

The waves are moving, but how is the water moving? Water is virtually incompressible, so as a wave passes and the water level falls, water must be displaced in one place as the wave becomes a trough, and replaced when it becomes a crest again. An ideal wave that is not breaking is closely represented by circular motion around axes parallel to the crests (Fig. 9.1). The diameter of the orbit is equal to the wave height at the surface, decreasing rapidly with depth and becoming a lateral oscillation on the bottom:

$$x = He^{-2\pi d/\lambda} \tag{9.3}$$

where x is the diameter of the orbit at depth d in a wave of height H and wavelength λ. At a depth of half a wavelength the orbital diameter is about 1/25 of the wave height – but this is still more than 80mm at a depth of 125m below a 2m swell with 250m wavelength. At the edge of the continental shelf, at 200m depth,

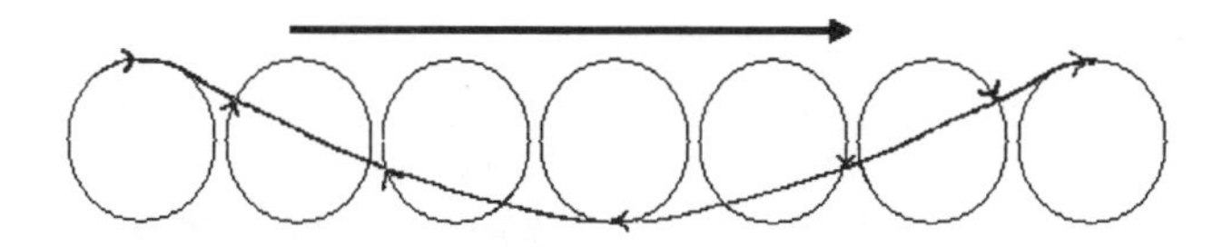

Figure 9.1 Wave motion.
A wave moves, but the motion within a wave in deep water is in circular, or near-circular, orbits.

the movement would be about 1 cm. *Wave base* is certainly not restricted to depths of about 10 m. A small movement repeated frequently over a long period not only has the ability to move particles but also to abrade them.

Once waves of stable dimensions have been generated by a storm (large λ, large velocity) they may achieve a velocity roughly 80 per cent of the wind's velocity, and when the wind dies, the waves are propagated from the area. The long-wavelength waves leave the short ones behind, and the short decay quickly. Waves lose height by about 1/3 (i.e. to 2/3) after travelling a distance of 5λ km. So waves that leave the storm with a height of 6 m and a wavelength of 250 m, will only be reduced to 4 m after a passage of 1250 km, 2·5 m after 2500 km, and so on. Such waves, which are called *swell* when they have left the storm that gave rise to them, can easily cross half an ocean. These waves travel on great circle paths, and are unaffected by coriolis forces because the water displacement is negligible.

Now, very large swell requires a long period of strong winds blowing steadily in direction and force over a considerable *fetch*, say, 3 days over 500–800 km. This is one reason for the huge seas in the Southern Ocean, where winds blow almost unobstructed from the west for much of the time. Tropical cyclones, on the other hand, have very strong winds (commonly reaching 120 km h^{-1}) but they blow in a pattern around a relatively limited area of about 200 km diameter. Waves therefore leave the cyclone in all directions, the largest being in the quadrant where the winds are blowing in the general direction of movement of the cyclone, the smallest (but still very large) in the opposite quadrant. Again, the larger waves leave the storm in all directions and become relatively harmless – and they indicate the approximate direction of the storm. Within the area of the storm, the seas are very dangerous mainly because of the confusion of wave trains with different directions. Wave trains are additive, as we saw in Figure 1.1.

If the swell or waves enter a current flowing in the opposite direction, the wave velocity is reduced, but the energy remains the same. The wavelength is shortened and the waves become higher (as any sailor knows who has sailed in the East Australian current, or the Agulhas of southeast Africa, with a stiff southerly wind). The effect seems to be quite out of proportion to the speed of the current (3–5 knots, 1·5–2·5 m s^{-1}). Throw a stone into a smooth part of a stream and see the effect.

When this swell reaches a continental shelf, it begins to "feel" the bottom – not much at first. A typical Pacific swell has a wavelength of 200–300 m and a period of 11–14 s; Atlantic swells are shorter, being typically 50–100 m with periods of 6–8 s. The outer continental shelf is not immune to disturbance on the sea floor due to swell because, as we have seen, there is still some movement at a depth of one wavelength. Storm waves of short wavelength have no effect at a depth of 200 m. As bottom friction slows the wave down, its wavelength shortens. The period remains the same. The energy of the wave is not reduced significantly near the surface so it becomes higher and steeper. This continues until the wave breaks. It breaks partly because it has become too steep, and partly because of the forward motion of the orbit at the crest of the wave, the water moving faster than the wave. By this time, the effect on the bottom is considerable and sediment may be stirred up.

As the wave enters shallow water, the physics changes and the velocity becomes a function of the depth of the water, d, rather than the wavelength: $(gd)^{1/2}$. There is a transition from the deep-water performance to the shallow-water performance, and the full expression for wave velocity is

$$c = \sqrt{(g\lambda/2\pi)\tanh(2\pi d/\lambda)}\ . \qquad (9.4)$$

Water depths, d, greater than about half a wavelength are considered to be deep because the hyperbolic term approaches 1 and Equation 9.1 applies. At water depths less than about $0{\cdot}05\lambda$ the hyperbolic term approaches $(2\pi d/\lambda)$ and the expression reduces to the shallow-water form above, $(gd)^{1/2}$. It is clear that the continental shelf is a zone of transition for the oceanic swell, and for much of it the full expression of Equation 9.4 should be used.

If the waves are coming in to the coast at an angle, the inshore part of a wave is slowed relative to the offshore part, and it is refracted. This affects the longer waves first. It is for this reason that waves come almost straight in to the beach whatever the wind direction, and also the reason that a boat sheltering behind a headland rises and falls to a gentle swell rather than to waves (except the locally generated waves). A slight net lateral movement of the sand owing to the swash of waves at a slight angle to the beach can result in massive sand transport over very short periods of time in a geological context.

The Polynesian navigators well understood the source and behaviour of waves, and were very skilled at detecting the various wave trains, and their refraction by islands.

Tsunami, seismic sea waves or "tidal waves"

Soon after Krakatoa, between Java and Sumatra, erupted in 1883, tidal gauges in Cape Town and Aden showed that a wave had crossed the Indian ocean at 725 km h^{-1} to Cape Town and at 560 km h^{-1} to Aden. In 1933, an earthquake in the Japanese trench sent a wave that reached San Francisco at 755 km h^{-1} (210 m s^{-1}). The latter seems to imply a wavelength of over 28 km, and a period of about 135 s, or 2·25 min (but both are usually much longer, wavelengths of 100–200 km with periods of up to 1 h or so, and heights of about 0·5 m in open ocean).

However, these waves travel at a velocity that is a function of water depth rather than wavelength – in other words, they behave as *shallow-water* waves

$$c = (gd)^{1/2}. \qquad (9.5)$$

Does that surprise you? It implies a mean water depth of 4140 m from Krakatoa to Cape Town, 2470 m from Krakatoa to Aden, and 4490 m across the Pacific. Once such a wave leaves the Japanese trench, a water depth greater than $\lambda/4$ is not found, so it behaves as a shallow-water wave. This has long been known, and tsunami velocities were used to estimate the mean depth of the oceans long before the oceans were surveyed.

10
ACOUSTICS: SOUND AND OTHER WAVES

Mechanical vibrations (i.e. elastic waves) with frequencies between about 50 Hz and 24 kHz can be detected by the human ear, and so constitute the audible range (much as there is a visible spectrum of colours in the range of electromagnetic radiation). The pure tones of music are very precisely related to frequencies, with middle C being 263 Hz.

A blow on a bell sets the bell vibrating. The part moving out compresses the air a little, and a wave is propagated; when it moves in, the air is rarefied a little and it too is propagated. Sound in air, water and other fluids (e.g. the outer core of the Earth), consists of compression–dilation waves. When these reach your ear, sympathetic movements are induced in your eardrum with exactly the same frequencies, and you hear the note belonging to that frequency. The amplitude of the wave is not maintained, mainly because energy is dispersing over the expanding surface of a sphere (nearly), which increases in area as the square of the distance from the source as it propagates from the source. So the wave is *attenuated*, and the sound gets weaker with distance from the source. You can *dampen* the sound by touching the bell.

Laplace showed that the speed or velocity of sound in a gas is given by

$$c = (\gamma p/\rho)^{1/2} \qquad (10.1)$$

where p is the pressure, ρ is the mass density of the gas, and this γ is the ratio of the specific heat capacity of the gas at constant pressure to its specific heat capacity at constant volume, c_p/c_v. This is strictly valid for ideal gases that obey Boyle's law (known as Mariotte's law in some countries). For gases with molecules consisting of only one type of atom, called monatomic gases, such as H_2 and O_2, $\gamma = 1{\cdot}67$; for diatomic gases, such as CO_2, $\gamma = 1{\cdot}41$. For polyatomic gases, the ratio is close to unity. For ideal gases, $p/\rho = RT$, where R is the gas constant and T is the absolute temperature; and the mass density is directly proportional to the pressure.

So the speed of sound in a gas is independent of pressure or density but proportional to the square root of its absolute temperature (from Equation 10.1 above). The speed of sound in air at 20°C is about $343\,m\,s^{-1}$.

In water, $\gamma \approx 1$, and the equation is usually written $c = (\gamma/B\rho)^{1/2}$ or $c = (K/\rho)^{1/2}$, where B is the compressibility of the water ($1/K$, with dimensions $M^{-1}LT^2$, inverse pressure and K is the bulk modulus)[1]. The bulk modulus of water is about 2·1 GPa (10^9 Pa), and its mass density about $1000\,kg\,m^{-3}$, so the speed of sound in fresh water is about $1450\,m\,s^{-1}$, being faster in warmer water than colder because the speed is inversely proportional to the square root of the mass density. Sound also propagates in solids, but in this case there are three types of wave: longitudinal (compression–dilation), transverse (shear), and surface waves. The compressional waves are fastest. The speed of sound – that is, the speed of the compression–dilation wave travelling longitudinally – is given by $(E/\rho)^{1/2}$, where E is Young's modulus of elasticity for the solid (see p. 80). Indeed, Young's modulus is measured in rods by timing the return of a sonic pulse. The speed of the shear wave is $(G/\rho)^{1/2}$. The shear modulus, G, of fluids is zero: shear waves cannot be transmitted in fluids.

Sound waves are reflected by solid surfaces – the echo. They travel at different speeds in different media and are therefore refracted much as light is refracted, and Snell's Law applies (see p. 36). Huygens' construction of wavefronts also applies (p. 38).

The *Doppler effect* has been noticed by all observant people. When a train passes, blowing its horn, the pitch or note changes as it passes, becoming lower as the train recedes. It would therefore be reasonable to assume that the note heard as the train approached was higher than the natural note of the horn that the driver and passengers hear, and that you would hear if both you and the horn were travelling with the same velocity in the same direction.

If the horn emits sound at a frequency of n Hz at velocity c while the train travels at a *relative* velocity $V\,m\,s^{-1}$ towards you, the horn emits n waves in 1 s, but they are contained not in a length c m, but in $(c - V)$ m. The pitch of the note is raised by a factor $c/(c - V)$. Conversely, as the train recedes, it is lowered by $c/(c + V)$. A horn that gives middle C on a train travelling at $25\,m\,s^{-1}$ ($90\,km\,h^{-1}$) will sound more like D flat when approaching a stationary observer and B sharp when receding. These are relative velocities; the ear does not know if it is stationary or not. The sound is transmitted at speed c in all directions, unaffected by the velocity of the source. (The *red shift* is also a Doppler effect, where lines in the spectra of stars that are receding from the Earth are shifted towards the red – a reduction of

1. The form $c = (K/\rho)^{1/2}$ is also valid for ideal gases, where K is the adiabatic compressibility, corresponding to the bulk modulus in solids ("adiabatic" means that heat neither enters nor leaves the system).

frequency, lengthening of wavelength. The light is also transmitted in all directions at speed c_0, unaffected by the velocity of the source.)

Elastic waves in the solids and porous solids of the Earth are of much longer wavelength than audible waves in air, but they must still be considered as sound waves. The frequencies of waves from earthquakes are principally in the range 10^{-3} Hz to 1 Hz. Consider the Earth. With the appropriate "listening" devices, there is a constant noise in the Earth punctuated, from time to time, by earthquakes of variable magnitude. There are four main types of wave generated by earthquakes, and they come under three headings, P, S, and L waves.

P waves oscillate in the direction of propagation by compression and dilation (you can think of them as push–pull waves). These waves are the same as sound waves transmitted through fluids – water and air, for example – but not the same as surface water waves. They are propagated at a speed or velocity of $[(K+4G/3)/\rho]^{1/2} = [(\lambda+2G)/\rho]^{1/2}$ and travel at different speeds in different media and so may be refracted.

S waves are shear waves that oscillate normal to the direction of propagation (like a violin string). These waves can only be propagated through solids because fluids are incapable of sustaining a shear stress. They are propagated at a speed of $[G/\rho]^{1/2}$, which is also different in different media, and so can be refracted. In fluids, G is sensibly zero, as we have noted. S waves travel at about half the speed of P waves.

In the context of the Earth, there is another set of waves called L waves that travel in the surface layers. They have very long periods and travel more slowly than P or S waves (remember, $c = \lambda/T$). There are two types of L wave. There are *Rayleigh* waves, in which the particles move in elliptical orbits; and *Love* waves, with horizontal oscillation normal to the direction of propagation. Surface waves exhibit *dispersion*, in that their velocity is a function of wavelength, much as sea waves are dispersed according to their wavelength.

These different waves travel with different speeds; and the speed of each wave is a function of the density, porosity, and elastic properties of the rock through which it passes. At surfaces of density contrast, a proportion of the energy of the wave is refracted, a proportion reflected. P and S waves can be identified on a seismograph and, as a first approximation, the difference in time between the arrival of the first P wave and the first S wave is proportional to the distance the waves have travelled from their source. This is the basis of earthquake location.

The severity of earthquakes is measured on two scales, the Richter scale and the Mercalli scale. The Richter scale is a measurement of the energy of the earthquake at the epicentre of the earthquake, its magnitude. The Mercalli scale is a measure of the effect of the earthquake, its local intensity, and it is an arbitrary scale rather like the Beaufort wind scale. The Mercalli scale runs from I (capital roman numerals), for those that are so weak that they are only detected by instruments, to XII where there is total destruction and visible waves at the surface. The Mercalli scale can be used to map the intensity of an earthquake over the area affected by it.

The Richter scale has been much modified since its introduction in 1935 for Californian earthquakes measured by stations with identical seismometers. Knowing the epicentre of the earthquake in California, the amplitude recorded on the seismometer was plotted against the distance of the station from the epicentre. Because the range of amplitudes was so large, the logarithm (to the base 10) of the amplitudes was taken, and this led to the comparison of the magnitude of earthquakes by considering the difference of the logarithms (to the base 10) of measured amplitudes – in particular, the comparison with the weakest. The modern Richter scale is an open-ended scale of general application that estimates the total energy of the source, E, from the magnitude, M, by using a formula of the form $\log E = a + bM$.

Although the scale is open ended (it is not a scale from 0 to 10), no earthquake has yet reached 9. What are the units of E? They are considered to be *ergs*, which are 100×10^{-9} joules. See Richter (1984) for a description of the scale that bears his name.

There is no convincing correlation between the Mercalli and Richter scales. Local geology has a very important influence on the local effects of earthquakes, as does the geology of the track from the epicentre.

See Telford et al. (1990: 140–62) for a more detailed exposition of seismic theory.

11
FLUIDS AND FLUID FLOW

Fluid flow is an important process in geology: it is also one of the least well understood, and the literature is full of errors. These errors arise mainly from ignorance and carelessness, as you would expect; but it is also a difficult subject. Its difficulty lies not so much in the mathematics involved as in the physical concepts.

General geology requires a good understanding of the principles of fluid flow for the various processes of sediment distribution (wind, water) and volcanic activity. Groundwater geology and petroleum geology require a good understanding of the physics involved in fluid flow, but it is one of the tragedies of science that some of the definitions are erroneously stated (even in the first two editions of the *AGI glossary of geology*). Ignore also the fluid flow part of Jaeger & Cooke's *Fundamentals of rock mechanics* (1979) in all three editions (I have a letter from Jaeger dated 26 April 1976 to this effect). The greatest tragedy is perhaps the erroneous or incomplete definition of the *darcy*, the unit of permeability in petroleum studies. This also shows some of the complexity of the concepts because the four authors of the papers defining the darcy were physicists and mathematicians.

Do not despair, even if your mathematics is weak, because it is not as difficult as all that. By the end of this chapter, the principles at least will be clear to you.

If you watch the smoke rising from a cigarette in a still room, it rises first in a vertical straight line and then suddenly breaks up into eddies and whorls. Similarly, a stream may flow with a smooth upper surface until it narrows, where it breaks in rough water. The smooth flow is called *laminar*; the rough, *turbulent*. Laminar flow is amenable to general mathematical treatment; turbulent flow is more difficult.

Consider water flowing in a channel of rectangular cross section and gentle slope. The component of weight down the slope provides the force that accelerates the water. Frictional resistance on the bottom and the sides, and the viscosity of

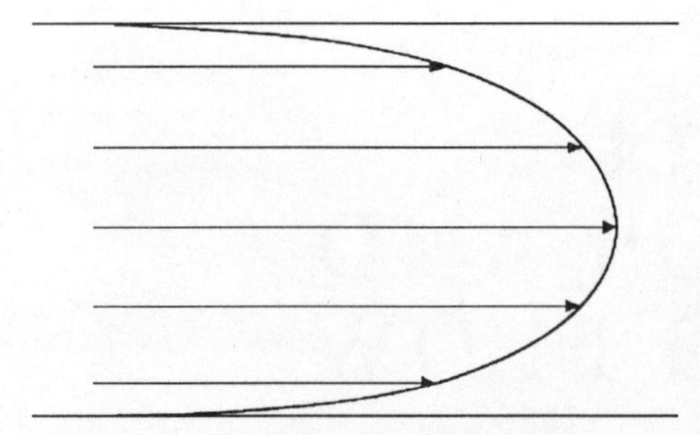

Figure 11.1 Speed of non-turbulent flowing water at the surface in a channel (in plan).
The vertical profile is similar to the bottom half of this diagram.

the water, provide the retarding forces; and when these forces have reached equilibrium, the water will have constant velocity – constant on all scales above the molecular if the flow is laminar, only in a gross sense if it is turbulent. The resistance at the sides means that there will be a velocity pattern in the surface water, with the greatest velocity at the centre, least at the sides, decreasing to zero in contact with the sides at what is called the *wetted surface* (Fig. 11.1). Similarly, there is a vertical velocity profile, from zero in contact with the base, to a maximum near the surface. (Why *near*, not *at*? Because there is air resistance to flow at the surface – less, certainly, than at the sides and bottom, but nevertheless a resistance.)

What then do we mean by *velocity* of a water stream when it is different in different positions in the water? In laminar flow we can say that the velocity is constant in any one *position* in the channel, but the variation of velocity *with* position means that we must define velocity of the channel water as a whole as the volumetric rate of flow divided by the normal cross-sectional area. Dimensionally it is $L^3T^{-1}/L^2 = LT^{-1}$. In units, it is $\mathrm{m^3\,s^{-1}\,m^{-2} = m\,s^{-1}}$.

Wind velocity also has a profile. That is why the stipulated height of anemometers is 10m above the ground.

Capillarity

The water of a fountain breaks up into drops that are nearly spherical; a tap drips drops that are nearly spherical; gas bubbles in soda water are nearly spherical; and you can remove a fly from water by dipping your finger on the fly and it is retained in the drop of water you extract. Some insects can walk on water, and if you look at them carefully you will see that their feet depress the surface slightly. You can fill a glass with water above the level of the rim. Capillarity is involved in all these things, and in the underground fluids, water, crude oil, and gas. Soils retain moisture because of capillarity; crude oil and natural gas reservoirs retain water because of capillarity.

The interface between air and water acts as if it were an elastic membrane in a state of tension, and it can support a pressure differential across it. This is *surface tension*, and it is defined in terms of the work required to separate unit area of the interface $[ML^2T^{-2}/L^2 = MT^{-2}]$. The same phenomenon exists between immiscible liquids or a gas and a liquid, and it is called *interfacial tension* (it is just a matter of terminology; the physics is the same). The cause of surface tension is molecular attraction at the interface. The reason a drop tends to become spherical is that the surface or interfacial tension tends to pull it into the shape that minimizes the surface area of the volume of water. For any liquid, surface tension is a constant for practical purposes, which is why many medicines are prescribed in units of drops.

If you insert a small glass tube (<5 mm diameter) into a pan of water (Fig. 11.2), the water rises in the tube – and it also rises against the glass of the tube both inside and outside. This is *capillarity*, and the shaped surface is called the *meniscus*. The surface of the water can be seen to be acting as a membrane in tension, and one would assume perhaps that the larger pressure is on the concave side. It is. The elevation of the meniscus is a function of the radius of the tube – the radius of curvature of the meniscus. Because the air pressure is sensibly the same on the water in the pan as on the water in the tube, the elevated water must be such that the pressure inside the tube level with the water level in the pan is atmospheric. That is, the elevated water has negative pressure: it is in tension.

$$p_c = -\rho g h_c = -p$$

where h_c is the elevation of the meniscus inside the tube above the water level outside. So the greater pressure is on the concave side of the meniscus.

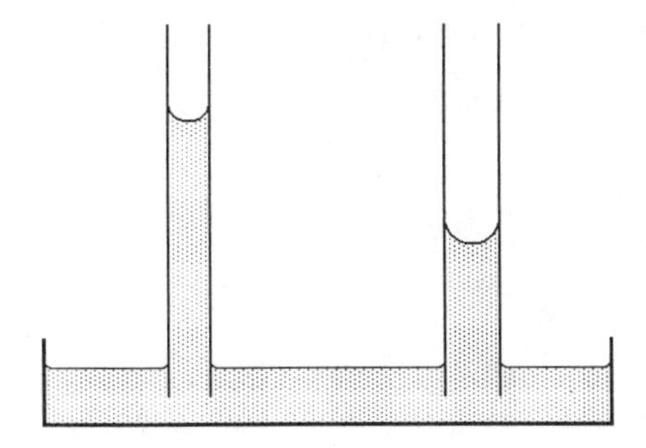

Figure 11.2 Capillarity.
The height to which water will rise in a narrow tube is a function of the internal diameter of the tube – the smaller the diameter, the higher the rise. In sedimentary rocks, the smaller the pores, the higher the capillary rise.

If you insert a glass tube into a pan of mercury, the reverse happens: the mercury level in the tube is below the level in the pan, and the meniscus is reversed, being higher in the middle. Again, the higher pressure is on the concave side of the meniscus. We therefore talk of *wetting* and *non-wetting* fluids (water being wetting on most materials and we refer to them as being "water-wet"). We are mostly concerned with wetting fluids.

Returning to the capillary aspects of water in tubes, the water rises in the tube until the upward component of capillary force is equal to the weight of the elevated fluid,

$$\rho g h_c \pi r^2 = \sigma 2\pi r \cos\varphi,$$

and

$$h_c = (2\sigma / r\gamma)\cos\varphi \qquad (11.1)$$

where σ is the surface tension $[MT^{-2}]$, γ the weight density, r the radius of the capillary tube, and φ is the contact angle (Fig. 11.3) ($\cos\varphi \approx 1$ for water on quartz).

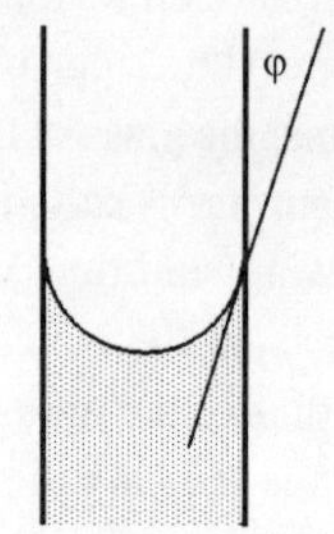

Figure 11.3 The contact angle φ.

If you fill a pipe with sand and stand it in a pan of water, water rises in the sand above the level in the pan, just as it did in the capillary tubes. It is essential for proper understanding to realize that this water in the sand is immobile. If it were not, we could drill a hole in the side of the pipe, let the water flow out and do work as it falls, and so have perpetual motion. This point was made by Pierre Perrault in 1674, and is just as true today.

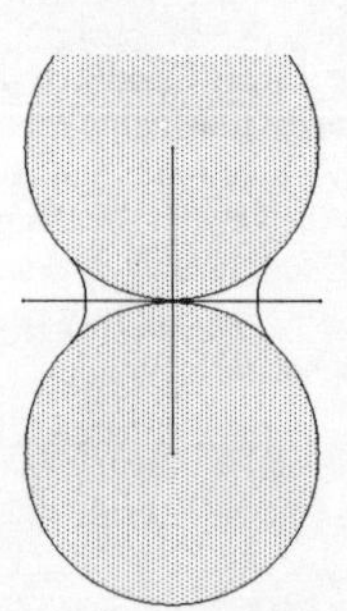

Figure 11.4 Ideal pendular ring, seen in section.
The water adheres to the solid surfaces around the points of contact. This water is immobile: it is not in hydraulic continuity with neighbouring pendular rings. The greater pressure is on the concave side of the interface.

If you put a measured quantity of water into a beaker, then put glass beads in up to the water level and hold them with a filter gauze, you will find that you cannot drain the same quantity of water from the beaker. Some is retained, and you can see it around the points of contact between the beads. This is called *pendular water*, and it is in pendular rings (Fig. 11.4). It is also the wetting phase. Clearly this water is immobile, not in hydraulic continuity, or it would not be there; and we presume that it is at a pressure rather less than that of the air in the pore space (because of the shape of the interface). It is held there by interfacial tension, and it would require energy to remove it – more energy than that required for the removal of the non-wetting phase. Precisely the same physics is involved in crude oil and gas reservoirs. The crude oil or gas displaced the water, but not all of it and pendular rings of water were left around the point contacts.

For pendular water around the point contact between two spherical particles, it can be shown that the capillary pressure is a function of the harmonic mean of the radius of curvature of the solids and the radius of curvature of the "meniscus":

$$p_c = \sigma r_{hm} = 2\sigma/[(1/r_s) + (1/r_w)],$$

where r_{hm} is the harmonic mean of the two radii.

This is a matter of great interest to geologists, but rocks are not strictly measurable in these terms. The contact angle is only constant with static fluids. If the fluid is flowing, the contact angle is greater for an advancing interface than a retreating one (examine a dew drop on a petal or leaf, or a tear on someone's cheek). Both surface tension and contact angle vary with water quality. It is not a topic, therefore, that is usefully approached quantitatively. We therefore approach the topic qualitatively.

When two immiscible liquids occupy the space of many pores, the non-wetting fluid occupies the space that minimizes its potential energy. This is the centre of the pore. The non-wetting fluid can only move if the energy of it is sufficient to move it past the constrictions between pores, which involves moving the interface past the constrictions. When the saturation of the non-wetting liquid is large enough for it to be continuous throughout the pore spaces, it can move fairly freely because there is no wetting fluid to displace from the pore passages.

Fluid statics

Equation 1.1, $p = \rho g z$, states that pressure in a fluid is proportional to depth and to the mass density of the fluid (ρ). We must first satisfy ourselves that this is true independently of the shape of the vessel or container.

The inability of fluids to resist shear stress leads to some fundamental propositions.

- *The free surface of a liquid in static equilibrium is horizontal.* This is a matter of common observation and practical utility in building and surveying. If the slope of the surface is not horizontal but has a slope θ, the weight (W) of a small element of liquid at this surface can be resolved into a normal component, $W \cos\theta$, and a shear component, $W \sin\theta$. The shear stress must be zero in a liquid, so θ must be zero.
- *Surfaces of equal pressure in static fluids are horizontal.* This really follows

from the first. If the fluid is not homogeneous and ρ is not a function of the depth, shear stresses will exist that will cause flow until they are eliminated, and ρ becomes a function of depth only. The shape of the container does not alter this.

- *The pressure at a point in static equilibrium is equal in all directions.* Consider a prism within the water that is so small that its weight is insignificant compared to its pressure (Fig. 11.5), and consider side b to have unit area. The area of side a is then $\sin\alpha$; and of side c, $\cos\alpha$. The force (pressure × area) exerted by the liquid on side *A* is then $p_a \sin\alpha$, and on side *C* it is $p_C \cos\alpha$. The force p_b on unit side b can be resolved into horizontal and vertical components, $p_b \sin\alpha$ and $p_b \cos\alpha$, respectively. Since the prism is in static equilibrium, the sum of the forces acting on it must be zero. Hence

  $p_a \sin\alpha - p_b \sin\alpha = 0$; and $p_a = p_b$,

  $p_c \cos\alpha - p_b \cos\alpha = 0$; and $p_c = p_b$,

  and the forces acting on the two parallel sides are clearly equal and opposite from the second proposition. Since $p_a = p_b = p_c$ is independent of the angle α, which may arbitrarily assigned a value and orientation, the pressure *at a point* in a static liquid is equal in all directions. Note carefully that the pressure on a submerged *object* is not equal in all directions.

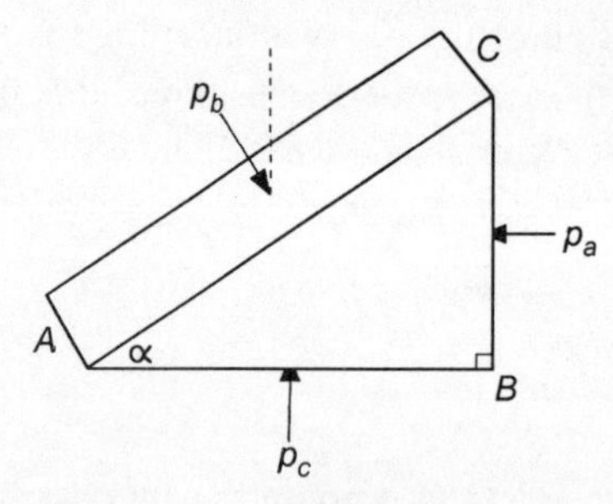

Figure 11.5 The forces acting on a small prism of water.

- *The horizontal force acting on a surface in a static fluid is the product of the pressure and the vertical projection of the area.* Consider the prism again. The force acting on side b, of unit area, is p_b, the horizontal component of which is p_b sin a. But sin α is the area of side a, which is the area of the projection of b onto a vertical surface. This proposition can also be shown to be true for curved surfaces by considering very small areas and their tangential slope.

The propositions above apply equally to the water filling the pore space of particulate solids, such as sand. If we now fill the containers with sand, it is clear that the water pressure in the pore spaces between the sand grains will follow the same relationship (bearing in mind that the grains will displace water and so raise the free upper surface). What about the pendular water? The short answer is that there is no water that behaves as pendular water in a single-phase liquid. It is the interfacial tension that makes the pendular water immobile, and this does not exist in a single-phase liquid.

What pressure does the solid–water mixture exert on the bottom of the container? and what about the solid component? What follows must be regarded as thought experiments because in small containers the pressure of particulate solids on the bottom does not increase once the thickness is greater than twice the diameter of the container.

Clearly the weight of the water-saturated sand exerts a pressure on the bottom of the container, and that total force is made up of a force due to the water and a force due to the solids. The fractional porosity f is the ratio of pore-water volume to total volume. The bulk weight is the sum of the weights of the water and the solids, $\rho_b gv$; and the bulk weight density is the bulk weight divided by the bulk volume, $\rho_b g$ $(=\gamma_b)$.

The total weight of the water-saturated sediment is made up of the weight of solids

$$W_s = \rho_s gv(1-f)$$

and the weight of the water in the pores

$$W_w = \rho_w gvf,$$

the sum of which is

$$\rho_b gv = W_s + W_w = [\rho_s(1-f) + \rho_w f]\, g\, v. \qquad (11.2a)$$

Buoyancy reduces the effective weight of the solids by the weight of water displaced. Should we not take this into account? No. If we do that we find that the total weight is less than the sum of the weights put into the container, unless the container was full of water to the brim before we put in the sand. In that case, the weight of water lost is equal to the weight of water displaced and this has to be taken into account.

If we divide Equation 11.2a by the area of the base, we get

$$\rho_b gh = \rho_s gh(1-f) + \rho_w ghf. \qquad (11.2b)$$

This suggests that the pressure exerted by the solids is only acting on the proportion of area occupied by solids $(1-f)$, and that the water is only acting on the proportion f.

Karl Terzaghi examined the question of buoyancy in concrete dams. He postulated that buoyancy only acts on the solid surfaces exposed to water in the pore spaces, but he found that the buoyant force was very close to $\gamma v(1-f)$, and that therefore the force of buoyancy acts on the total area $(1-f)$, irrespective of the areas of contact between grains. He later found the same results in clays. Consequently, he stated that the total stress in porous solids is divided between the *effective stress*, σ, transmitted through solids and the pore-fluid pressure (he called it the neutral stress). To be a bit more precise,

$$S = \sigma + p, \tag{11.3}$$

where S is the vertical component of total stress. This is known as Terzaghi's relationship. It is now known not to be *exact*, but it is sufficiently close for most geological purposes. (One cannot avoid a clash of symbols; this σ is not the surface tension.)

Reverting now to Equation 11.2a we see that that is but one possible partition. Substituting into Equation 11.3 the expressions for S and p, we can write

$$\begin{aligned}\sigma &= S - p = \rho_b gh - \rho_w gh \\ &= [\rho_s(1-f) + \rho_w f - \rho_w]\, gh \\ &= [\rho_s(1-f) - \rho_w(1-f)]\, gh.\end{aligned}$$

The effective stress is due to the partial density of the solids less the water they displace, and it is this partition of total stress that agrees with experimental results (see Hubbert & Rubey 1959: 139–42).

The result is important because it is the effective stress that leads to mechanical compaction of sediments and sedimentary rocks under gravity. Terzaghi called the water pressure the neutral stress because it has no direct rôle in the deformation of the grains.

Another important result is that the depth of water over the top of the sand makes no difference to the *effective* stress. Imagine that there is a depth l of water above a vertical thickness h of water-saturated sand:

$$\sigma = \rho_b gh + \rho_w gl - \rho_w g(h + l) = (\rho_b - \rho_w)\, gh.$$

The depth of water does, of course, contribute to the total stress, S.

Reynolds numbers

Osborne Reynolds (1842–1912) carried out some elegant experiments on the flow of water in pipes in the 1880s, and was interested in the conditions that separated the stable from the unstable, the laminar from the turbulent. He found that it was a function of the internal diameter of the pipe and the velocity and viscosity of the water in it, but argued that since *units* are arbitrary, there must be a dimensionless relationship. Noting that the dimensions of *kinematic* viscosity ($\nu = \eta/\rho$) are those of a length multiplied by a velocity, he took as his dimensionless number dV/ν, where d is the diameter of the pipe and V is the velocity of the liquid (volume rate per unit area normal to flow, $L^3T^{-1}L^{-2} = LT^{-1}$). This is now called a Reynolds number. It consists of a characteristic length multiplied by a characteristic velocity divided by the kinematic viscosity of the fluid. It has the indefinite article because there are different ways of defining the components. (Reynolds actually took the inverse of this but it was changed to the form given, presumably to give a large number rather than a small fraction, and that is how it is now defined.)

In pipe and channel flow, the characteristic length is the *hydraulic radius* (see below). What should it be in porous rocks? It is common practice to take the mean grain size, but this is a very indirect quantity because it is precisely the part of the rock that fluids do not flow through. We shall return to this problem when we consider water flow through porous rocks.

Hydraulic radius

With pipe flow, there is no difficulty in measuring the radius. But what if the pipe were square or elliptical? or if the pipe were not quite full?

It has been found that the unifying parameter is obtained by taking the volume of water and dividing it by its wetted surface (if the dimensions remain constant, you can take the cross-sectional area divided by the wetted perimeter). This is called the hydraulic radius, and it has the dimension of length. The hydraulic radius of a pipe of circular cross section and constant internal diameter d is

$$R = \pi(d^2/4)l/(\pi d l) = d/4. \qquad (11.4)$$

If we never used anything but pipes that were full, this would not be necessary. But pipes are not always full, and water can flow in channels as well as pipes.

The hydraulic radius is perhaps best understood with channel flow. Consider unit length of water in laminar flow down a gently inclined channel of width w, the depth of the water normal to the bottom being h. Its volume is wh, and its wetted surface is $w + 2h$. The forces acting on this volume of water are gravitational (the component of weight down the slope θ) and frictional (the resistance of the wetted surface). With laminar flow, these are equal and opposite, so

$$wh\rho g \sin\theta - \tau_0(w + 2h) = 0 \qquad (11.5)$$

where τ_0 is the boundary shear stress. So the ratio of boundary shear stress to the component of weight of unit volume down the slope is

$$\tau_0/\rho g \sin\theta = wh/(w + 2h). \qquad (11.6)$$

This is the volume divided by the wetted surface area – the hydraulic radius. The hydraulic radius is a measure of both size and shape.

It also applies to rivers, where it is called the *hydraulic depth* because the right-hand side of Equation 11.6 approaches h as w becomes large compared to h.

Solids settling in static fluid

If you place a small pebble at the surface of some static water and let it fall, it will accelerate until the frictional resistance of the water equals the weight the pebble. Its velocity then will be constant at what is known as its *terminal velocity*. Likewise, if you jump out of an aeroplane without a parachute, the speed at which you hit the ground will not be the same for all small heights, but once high enough to achieve your terminal velocity (about 200 $km\,h^{-1}$ if you spread your arms and legs, but nearer 240 $km\,h^{-1}$ if you go into a dive with your arms like swept-back wings), that will be the speed at which you hit the ground from any greater height (and it will be terminal!).

The larger the solid object, the larger the *sympathetic* flow that will be generated by its passage through the water. The more solid objects there are, the larger will be the sympathetic flow. The *shape* also affects the speed at which the object falls. During the Second World War, a special bomb was designed to fall faster than the normal bomb so that it would penetrate deeper into the ground before exploding (it was called the *earthquake* bomb, and the idea was to shake down bridges and buildings; it worked). The other extreme is the parachute.

Dimensional analysis can be used to find the form of the equation giving the terminal velocity of a single *small* solid sphere:

$$V = \{[(\rho_s/\rho_w)-1]^a\, g^a\, d^{a+1}/b\nu\}^{1/(2a-1)}, \tag{11.7}$$

where V is the terminal velocity of a *small* sphere of diameter d and mass density ρ_s falling through static fluid of mass density ρ_w and kinematic viscosity ν. The dimensionless numbers a and b must be determined experimentally. When $a = 1$ and $b = 18$, we have Stokes' Law. Stokes' Law applies *only* to a single small sphere and then only when its terminal velocity is small enough for terms in V^2 to be neglected. It should not be used for multiple grains settling in a fluid. The point here is that when V^2 is not negligible, kinetic energy is not negligible and the relationship between V and ρ is no longer linear.

It has been found that Equation 11.7 applies reasonably well to multiple grains, and grains large enough to give significant terms in V^2 – that is, it can be used for sediment settling provided the coefficients a and b are determined for similar material.

It was said above that resistance to a moving object is proportional to the square of its velocity, independent of shape. If an object is falling in air, it cannot "know" whether it is falling, or being suspended by a rising wind.

Winnowing

In many parts of the world, wheat is separated from the chaff by beating the corn with a stick and then tossing the wheat and chaff in a breeze. The grains fall back into the basket; the chaff blows away. The reason for this is that the grains are aerodynamically better shaped than the chaff (like the parachutist and his parachute).

Volcanic ash and dust

When a volcanic eruption, such as that of Mount St Helens on 18 May 1980, throws ash and dust into the air, the energy that elevates the fine dust is kinetic and thermal, and this combination may be large enough to put material into the stratosphere (about 10–50 km altitude). The ash plume of Mount St Helens reached an altitude of 25 km, and ash and tephra were spread out 100 km to the northeast. The initial velocity required to put a mass of 10^{-3} kg to an altitude of 25 km by kinetic energy alone, ignoring friction, can be found by equating kinetic energy to the potential energy of the mass at an elevation of 25 km:

$$mV^2/2 = mgh$$

and

$$V = (2gh)^{1/2} = (19{\cdot}6h)^{1/2}\,\mathrm{m\,s^{-1}}. \tag{11.8}$$

The initial velocity required to raise a mass to 25 km would therefore be at least 700 m s^{-1} – about twice the speed of sound – and this would be independent of the mass of the piece in the absence of friction. It is therefore probably safe to conclude that only dust-size particles reached 25 km, and that they were carried there largely by thermal convection currents generated by the heat of the volcano. If 100 m s^{-1} is a likely maximum launch velocity, 500 m is a likely maximum altitude from kinetic energy alone (if 200 m s^{-1}, 2000 m).

The velocity at which the material falls to Earth can be seen from Equation 11.7 to be proportional to its size and to its density. In general, the larger, denser fragments will fall back to Earth sooner than the smaller, less dense. We must not expect Stokes' Law to prevail because (a) some of the terminal velocities may be large enough for the kinetic-energy term in V^2 to be significant and (b) there will be interactions between the mass of fragments falling at different speeds.

When Krakatoa (northwest end of Java) erupted in 1883, a cloud of fine dust spread around the Earth and affected sunrises and sunsets for several months – as happened after the eruption of Pinatubo in the Philippines in June 1991. This material must have been very fine; but if the terminal velocity is very small, an air current rising at the same velocity would suspend all such particles (indeed, it was probably carried to these heights in a rising column of warm air or gas). This may happen with rain drops in a thunderstorm, leading to a "cloudburst" when the uplift ceases or when the water saturation reaches some critical value.

Solids in flowing fluid

In liquids

The murky colour of river water after heavy rains is due to the sediment carried in the water. Part of the total sediment load will be in *suspension*; part, in *traction* along the river bed. It will be clear that, although we cannot apply Stokes' Law to this problem, the principles are much the same. The finer, less dense, material will be in suspension; the coarser, denser, will be in traction; and there will be a critical size and density above which it will not move. There may well be a critical Reynolds number at which particles of a particular size will move.

The physical principle involved is that grains will move on account of the energy of the fluid until they arrive in a position in which the energy is insufficient to move them farther. The energy of the water is partly transferred to the solids by virtue of the frictional resistance at the water/solid interface.

In gases

Windblown sand Once again, we cannot expect Stokes' Law to apply to windblown sand, but we can expect the principles to apply. The distinction between sediment transport in water and in air lies mainly in the smaller mass density of air, and its smaller viscosity. Equation 11.7 suggests that terminal velocity is inversely proportional to the fluid mass density and to its viscosity. So we would expect water to be able to move much larger grains than wind, if their velocities are equal. Buoyancy also plays a part. As the dimensionless ratio ρ_s/ρ_w decreases and approaches 1, so the terminal velocity decreases and the ease of movement by the fluid increases.

See Bagnold (1941) for a classic, but rigorous treatment.

Nuées ardentes, avalanches and turbidity currents When a volcano erupts, fine hot powder in hot gases may cause a plume to rise to great altitudes, as it did over Mount St Helens in the USA in May 1980. If the amount of solids exceeds the buoyant effect of hot gases, the cloud descends from the volcano as *nuées ardentes*, as happened on Mont Pelée in Martinique, West Indies, in May 1902 with the destruction of the town of St Pierre and all but one or two of the town's 30 000 inhabitants.

When solids are suspended in a fluid, that fluid acquires a greater bulk density, and so tends to move to the most stable position it can reach, minimizing the potential energy of the system. Snow avalanches and turbidity currents are similar.

Bernoulli's theorem

Water can be regarded for practical purposes as incompressible over the pressure changes that may take place in relatively short distances and with slow movement. Daniel Bernoulli (Switzerland) used this and the principle of conservation of energy to formulate an energy balance.

Consider an ideal liquid flowing through a pipe of decreasing diameter (Fig. 11.6). The work done on the liquid in the pipe is equal to the change of energy of the liquid. At station 1, the force applied to the liquid is equal to $p_1 A_1$, where p is the pressure and A is the area normal to the flow; and in moving the liquid a small distance δl_1 in the small interval of time δt, the work done is equal to $p_1 A_1 \delta l_1$. Likewise, the work done at station 2 is $p_2 A_2 \delta l_2$. The weight of liquid passing station 1 during the interval of time δt is $\rho g A_1 \delta l_1$; and the same weight of liquid leaves station 2.

The *potential* energy E_p per unit of mass entering at station 1 is gz_1, so the potential energy of unit weight is z_1 and the potential energy of the liquid entering at station 1 is $\rho g A_1 \delta l_1 z_1$, and that leaving at station 2 is $\rho g A_2 \delta l_2 z_2$.

The *kinetic* energy E_k per unit of weight is $q^2/2g$, where q is the volumetric flow rate; so the kinetic energy of the liquid entering at station 1 is $\rho g A_1 \delta l_1 q_1^2/2g$; and

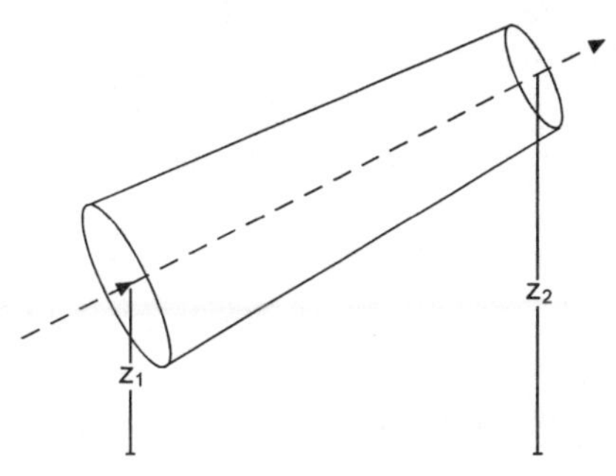

Figure 11.6 Bernoulli's theorem relates the energy changes in flow through a converging tube.

that leaving at station 2 is $\rho g A_2 \delta l_2 q_2^2/2g$.

There is also thermal energy E_t, but we are considering an ideal liquid of zero viscosity.

The principle of the conservation of energy requires that the sum of the work done, potential energy, and kinetic energy shall be constant:

$$p_1 A_1 \delta l_1 + \rho g A_1 \delta l_1 z_1 + \rho g A_1 \delta l_1 q_1^2/2g = p_2 A_2 \delta l_2 + \rho g A_2 \delta l_2 z_2 + \rho g A_2 \delta l_2 q_2^2/2g$$

in which each term has the dimensions of energy, ML^2T^{-2}. Since $\rho g A_1 \delta l_1 = \rho g A_2 \delta l_2$,

$$(p_1/\rho g) + z_1 + (q_1^2/2g) = (p_2/\rho g) + z_2 + (q_1^2/2g)$$

or,

$$(p/\rho g) + z + (q^2/2g) = \text{constant}. \tag{11.9}$$

This is known as Bernoulli's theorem, and it is strictly valid only for incompressible, frictionless liquids. Each term has the dimension of *length*. What meaning is to be attached to the velocity, q? The velocity of a particle of water depends on its position. It is slow near the fluid container, faster near the centre. As before, we take the velocity to be the volumetric rate of flow divided by the cross-sectional area of flow: $L^3 T^{-1} L^{-2} = LT^{-1}$.

So what do we really mean by the kinetic energy of water flowing in a pipe or channel? A particle of mass m and velocity V has kinetic energy $mV^2/2$, or, per unit of weight, $V^2/2g$. But water flowing in a pipe or channel has a profile in three dimensions, so what is its total kinetic energy? If we know the profiles accurately, we can obtain the total kinetic energy; but we can rarely do that. So we write

$$E_k = aV^2/2g, \tag{11.10}$$

where a is a factor that takes the total water flow into account. For laminar flow in circular pipes, it turns out that $a = 2$. For turbulent flow, a is a little more than 1.

The first term in Equation 11.9, $p/\rho g$, is known as the *pressure head* (see Fig. 11.7), and it is the vertical height of a column of liquid of mass density ρ that can be supported by a pressure p. The second term, z, is merely the elevation of the element of liquid above some arbitrary horizontal datum plane (strictly a surface of constant potential energy). It is called the *elevation head* (less desirably, the *potential head*). The third term, $q^2/2g$, is called the *velocity head*. The sum of these is called the *total head*, and with an ideal, incompressible liquid of zero viscosity, the total head would remain constant. In fluid flow in rocks the velocity head is negligibly small and is neglected. The total head is proportional to the energy of the water (lacking only the factor g).

Real liquids are not ideal, and energy will be lost to heat, leading to a loss of total head as the real liquid moves.

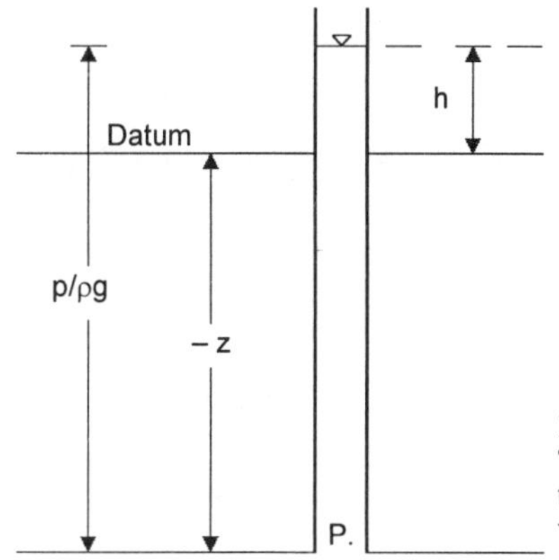

Figure 11.7 Heads.
The pressure head ($p/\rho g$), elevation head (z, negative when below the datum), and total head (h), which is the algebraic sum of the other two.

Fluid flow in rocks

Henry Darcy (note the spelling, Henry not Henri, no apostrophe in Darcy) was an engineer responsible for water supply in the town of Dijon in France, and he conducted experiments to determine the effect of sand filters on the flow of water through pipes. He published the results in 1856.

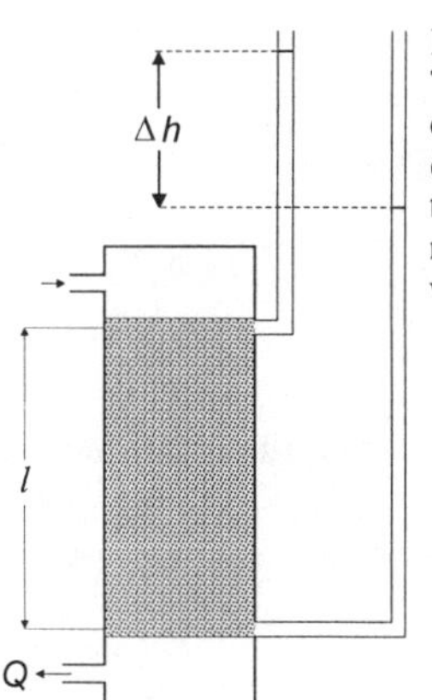

Figure 11.8 Diagram of Darcy's apparatus (in section).
The sand is packed in the cylinder over the length l, and the water energy lost to friction by flow through the sand is proportional to the difference in elevation of the water surfaces in the manometers. If the flow were reversed, the levels in the manometers would be reversed, but the value of Δh would be the same. If the apparatus were to be inclined, the results would still be the same.

Darcy filled a cylinder with sand, supported on a fine mesh, in apparatus shown diagrammatically in Figure 11.8. Manometers were placed near the top and bottom of the sand, and water was passed through the sand in the cylinder. The volumetric rate of flow was measured, as were the elevations of the water in the manometers (Fig. 11.9). Darcy observed that "for sands of the same nature" the flow was proportional to the difference in elevation of the levels in the mano-

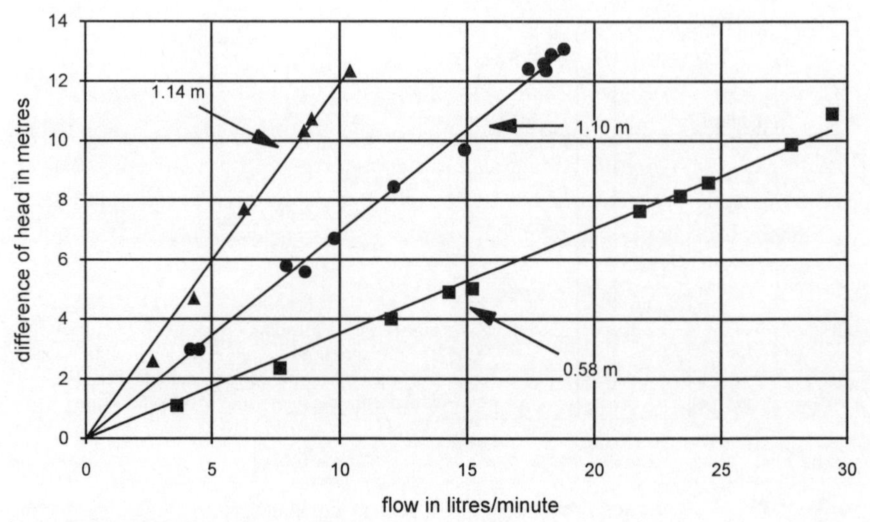

Figure 11.9 Darcy's results.
These are plotted as the difference of head in metres due to flow at the rates given in litres per minute through filters of different sands and different lengths (marked). The straight lines are the regression lines constrained to pass through the origin – no difference of head when there is no flow.

meters, and to the area normal to flow, and inversely proportional to the length of the sand,

$$Q = KA\Delta h/l. \tag{11.11}$$

where Q is the volumetric rate of flow, A is the cross-sectional area of the cylinder, Δh is the difference of the water levels in the manometer (total head), l is the macroscopic length of flow in the sand, and K is the proportionality constant called the *coefficient of permeability* or *hydraulic conductivity*.

It is important to understand from the outset that the statement "fluid flows from high pressure to low pressure" is false. The water in a swimming pool does not flow from the bottom to the top. The water in an artesian basin actually flows from low pressure to high pressure. Darcy was not measuring the difference of *pressure* across the sand; he was measuring the difference of *energy* as measured by the elevations in the manometers (see Bernoulli's theorem, Eq. 11.9). This is the difference of total head, h.

We now rewrite Equation 11.11 as

$$q = Q/A = K\Delta h/l, \tag{11.12}$$

where q is the *specific discharge* (or *approach velocity*, because it is the real velocity of the water in the cylinder just above the sand), and it has the dimensions of a velocity $[LT^{-1}]$. The term, $\Delta h/l$, called the *hydraulic gradient*, is dimensionless; so the coefficient of permeability has the dimensions of a velocity and is not therefore a true constant but a material constant that varies with both the properties of the fluid and the pore characteristics.

The coefficient of permeability, K, clearly takes several factors into account. Darcy found different values for K for different sands. Had he used a liquid other than water, with a different viscosity, no doubt K would have had different values again. And the hydraulic gradient is clearly an energy term for unit weight of the liquid, lacking the factor g. So K must include the acceleration due to gravity, g, and the mass density of the liquid, ρ. Dimensional considerations lead to

$$K = k\rho g/\eta \qquad (11.13)$$

where η is the coefficient of viscosity, and k is the permeability attributable to the permeable material alone, called the *intrinsic permeability*. But k has the dimensions of an area, so it too is a material constant (as you would expect), and must embrace various attributes of the sand.

The size of the pores clearly affects water flow through them, so we would expect the *hydraulic radius* [L] to play a part in intrinsic permeability. Hydraulic radius, as we have seen, is a measure of both size and shape of the passages for the water flow. It is the characteristic dimension of the pores, but must not be regarded as an average dimension: it is just the volume of pore space available for water flow divided by the bounding area. The hydraulic radius is the same (with but statistical variation) for $1\,m^3$ as for $10\,m^3$. Dimensions indicate that intrinsic permeability is proportional to the square of the hydraulic radius.

Water flowing through the pores of the sand does not flow in a straight line over the distance l in Figure 11.8 but takes a much longer path the mean length of which for all particles of water is l_t. This aspect of fluid flow through porous media is called *tortuosity*, defined l_t/l. It is dimensionless; and it is a vector in the sense that the value of tortuosity in sedimentary rocks usually varies with direction, being much larger across the bedding than parallel to it. Tortuosity is the only component of permeability in sedimentary rocks that varies with direction, and therefore must account for anisotropy of permeability in rocks.

We conclude, therefore, that intrinsic permeability includes tortuosity and the square of the hydraulic radius. For a more detailed analysis in terms of measurable rock parameters, see Chapman (1981: 49*f)*. For a more detailed mathematical analysis, see Hubbert (1940, 1969).

Finally, there is the matter of the energy of water. The total head, h, as we have seen, is an energy per unit of weight of the liquid; the term gh is an energy per unit of mass of the liquid. The *potential* of a fluid is its energy per unit of mass, gh, and has the symbol Φ. So,

$$\Phi = gh = gz + p/\rho.$$

The potential gradient is therefore $g\Delta h/l$.

Important groundwater definitions

(If you find definitions at variance with these, examine the physics carefully to determine which is correct.)

artesian Wells or aquifers in which the energy of the water is sufficient to raise it above the level of the ground without artificial aid are said to be artesian. An aquifer is artesian if its total head is greater than the elevation of the land surface above the same datum.

equipotential line or surface A line or surface on which all points of equal energy, equal fluid potential, lie. See **potential**.

head Refers to the vertical length of a column of fluid, usually liquid. It is an energy per unit of weight, with dimensions of length, not pressure. Ambiguous when not qualified. See Figure 11.7.

elevation head or simply **elevation** The potential energy per unit of weight due to elevation, i.e. the elevation of the point at which pressure, p, is measured, above (+) or below (–) an arbitrary horizontal datum. $h_e = z$.

pressure head The vertical length of a column of liquid supported, or capable of being supported, by pressure p at a point in that liquid. $h_p = p/\rho g$.

velocity head Head due to the kinetic energy of the liquid. Negligibly small in most geological contexts. $h_v = V^2/2g$

total head The algebraic sum of the elevation and pressure heads (neglecting velocity head). It is proportional, lacking only the factor g, to the energy of the liquid at the point where the pressure p was measured. $h = z + p/\rho g$. See **potential**.

static head The pressure head in a borehole that is not producing. See **static level**.

hydraulic conductivity See **permeability**.

hydraulic gradient The difference of **total head** divided by the macroscopic length of porous material between the points where the total head is measured. $\Delta h/l$ (dimensionless). Properly it is the steepest gradient at a point in the fluid. Note that the gradient of a **potentiometric surface** is not strictly the hydraulic gradient unless the aquifer is horizontal (but it may be a sufficiently close approximation). Hydraulic gradient is not the same thing as pressure gradient.

hydraulic radius A measure of the size and shape of the space through which liquid flows, it is the volume of liquid divided by its wetted surface area. Dimension $[L]$.

permeability The quality of rocks that allows the passage of a fluid; a measure of the ease of passage of fluids. There are two measures of permeability:

coefficient of permeability or *hydraulic conductivity* The coefficient K in Darcy's law when written $q = K\Delta h/l$. It takes into account both the fluid properties and the properties of the permeable medium. Dimensions of a velocity, LT^{-1}.

intrinsic permeability A measure of the material properties allowing the flow of fluids, usual symbol, k. Dimensions of an area, L^2. The coefficient of permeability is related to intrinsic permeability by $K = k\rho g/\eta$.

potential The energy per unit of mass of the liquid at a point where the pressure p is measured. $\Phi = gh = gz + p/\rho$. Dimensions, L^2T^{-2}.

potential gradient The loss of energy per unit of length along the macroscopic flow path. $g\Delta h/l$. Dimensions of an acceleration, LT^{-2}. See **hydraulic gradient** above.

potentiometric surface The surface obtained by mapping the total head of an aquifer. The potentiometric surface can be contoured (equipotential lines) and fluid flow is in the direction normal to these contours, down the slope of the surface in the direction of decreasing total head or potential. The synonym *piezometric surface* is not to be preferred because it is a matter of potential, not pressure.

static level The level of the water in a well that is not producing, and has stabilized. This level may be given relative to the ground surface or the bottom of the well; but if given relative to some *horizontal* datum surface becomes synonymous with **total head**.

water table In practice, the free water surface of an unconfined aquifer in a well. Strictly, it is the level *in* the zone of saturation at which the pressure is atmospheric, i.e. water raised in rocks by capillary action is ignored.

12
SOME DANGERS OF MATHEMATICAL STATISTICS

Mathematical statistics and statistical methods pervade modern science, but there are great dangers in using techniques that you do not thoroughly understand. Statistics is a background consideration of the most common geological operations, such as sample taking, fossil collecting, and estimating the proportions of different constituents in rocks. Is the sample you have just taken representative? You went for the freshest part of the outcrop – the freshest part that you could reach – so really it is most *unlikely* to be representative. But if you don't have a fresh sample, you may not know sufficiently well what the rock really is or was!

The fundamental problem with statistics in the physical and natural sciences lies in the concept of *randomness*. When you toss a coin, it is reasonable to assume that it is a matter of pure chance whether it comes to rest head or tails upwards – each is equally likely if the spin of the coin is fast. If large meteorites hit the Earth from time to time, it would be a reasonable assumption that the rotation of the Earth on its axis and around the Sun, coupled with the indeterminacy of the positions of the larger meteorites, means that each square kilometre of the Earth is as likely to receive an impact as each other. If that is found not to be the case, then one conclusion (there may be others) will be that meteorites do not approach us from *all* directions. If they lie in the plane of the solar system or ecliptic (as they do, ±30°), impacts will be more likely on a square kilometre in the tropics than at the poles. Already the idea of meteorites randomly distributed in space would have to be qualified.

Then there is the probability of such an impact being discovered. This varies from near certainty if the meteorite hits an inhabited area, to a rather small probability if it lands in the Southern Ocean at night in winter. So only the land areas can be considered. The larger northern hemisphere land area would have to be taken into account.

In a thin section of rock you can measure the areas of the different components with great precision, but is the thin section itself representative? What would have been the result if it had been orientated normal to the one you have? What if you had been able to get it from a metre or two inside the rock?

The assumption is that the areas in a thin section are proportional to the volumes in the whole sample, and that the sample is representative of the whole. This applies to porosity as well. Randomness comes into this assumption because of the assumption that the slide is representative of the whole.

Linear regression

Linear regression analysis is the search for the best-fitting straight line to two sets of data that are believed to correlate, so that the value of one can be estimated if the value of the other is known. Such relationships are particularly useful if one quantity is easy to measure, the other not. Linear regression equations are in the form $y = mx + b$, where x is the independent variable, m is the slope of the line, b is the y intercept (the value of y when $x = 0$) and y is the dependent variable. This line always passes through the point (x, y) representing the mean values of x and y. The data points fall on the straight line only when there is a functional linear relationship. Normally there is a scatter due to errors of measurement, or disturbances due to causes not taken into account. The amount of scatter is measured by the correlation coefficient, r or r^2, and there are tests to determine if a particular value of r could reasonably have arisen by chance. The value of r lies between -1 and 1, -1 indicating perfect negative correlation, 1 perfect positive correlation, and 0 no correlation at all. Do not make the mistake of thinking that correlation coefficients close to 1 or -1 are necessarily significant. The significance depends also on the number of *degrees of freedom*. Two pairs of figures, two points on a chart, will always have a correlation coefficient equal to 1 or -1; but there are no degrees of freedom, if you think about it, so it is not significant. There may or may not be correlation. A minimum of three points is required for any assessment of significance. On the other hand, a correlation coefficient of 0·2 with 1000 degrees of freedom is very significant. (See Note 4 if you are unfamiliar with the concept of degrees of freedom, and significance.)

One other point: if you determine the slope and the intercept in the formula $y = mx + b$, and want to make x the dependent variable, you must not simply rearrange it into $x = (y - b)/m$. Another formula has to be computed unless $r^2 = 1$. The first, $y = mx + b$, is called the regression of y on x, and the square of the error in y is minimized: the second, $x = m'y + b'$, that of x on y, and the square of the error in x is minimized.

Figure 11.9 contains the results of the first two series of experiments conducted by Darcy, and the series conducted independently by Ritter, and reported by Darcy. It is fairly clear that the lines representing the best fit of the data are good, but not perfect, and that the relationship is indeed linear, as Darcy's equation (Eq. 11.12) indicates. The best check on linearity is to do what Reynolds of Reynolds number fame did: plot the logarithms of the data (natural or base 10) or compute the linear regression of the logarithms, and the *slope of the line will indicate the order of the association.* (Zeros and negative numbers in the data can be eliminated by adding a constant to all the data.) For linear relationships, $y = bx^m$, the slope m should be 1 and $\ln y = \ln x + \ln b$. If the slope is far from 1, there is little point in proceeding with the *linear* regression analysis, but the order of the association will be evident from the logarithmic data. The logarithms of Darcy's data are plotted in Figure 12.1, and it is clear that although the slopes are not all exactly 1 (marked by the dashed line), they are close to 1 and we can regard Darcy's equation as being linear.

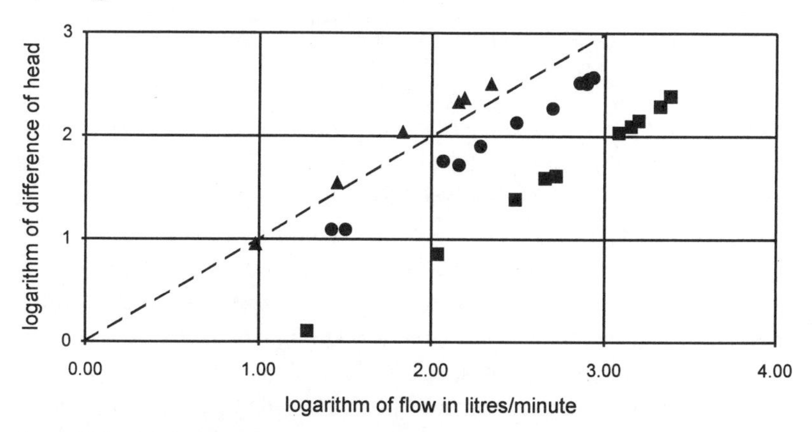

Figure 12.1 Darcy's data, plotted as logarithms.
The slopes are close to 1 (marked by the dashed line) indicating a linear relationship between the difference of head, Δh, and the discharge, Q.

Table 12.1 Data for the correlation example.

x	y
1	0·5
1·5	1·13
2	2·00
2·5	3·13
3	4·50
3·5	6·13
4	8·00
4·5	10·13
5	12·5

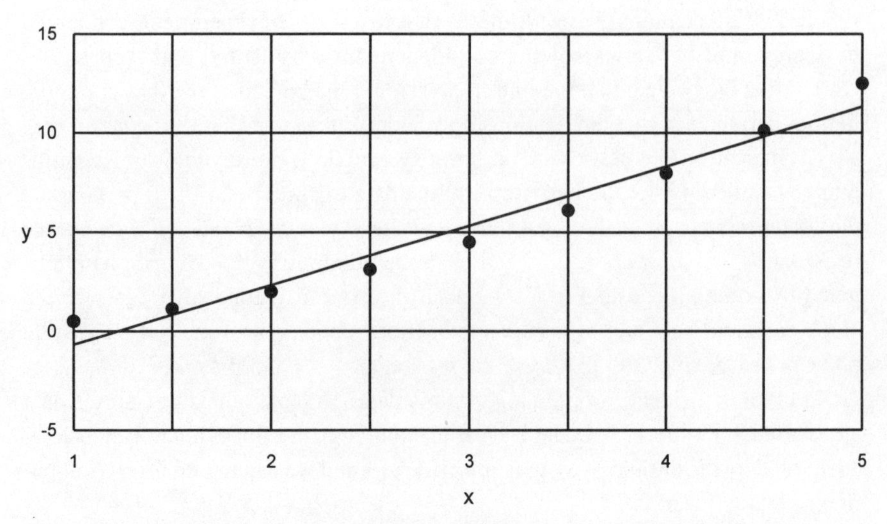

Figure 12.2 Plot of the data listed in Table 12.1, with the linear regression line.
The systematic differences are evident although statistical analysis indicates a very significant linear relationship that implies a random error about the regression line.

The data in Table 12.1 were obtained experimentally. Using a pocket calculator, it is found that the linear regression equation is $y = 3x–3{\cdot}66$ and the correlation coefficient, r, is 0·97. This coefficient is extremely significant (Student's t, which is one method of assessing significance, is 14·0 for 7 degrees of freedom – two points will always lie on a straight line – whereas there is a probability of about one per cent or 0·01 that t will be 3·5 or larger by chance, and perhaps a million to one that it will be as large as 14 or larger). You might therefore have great confidence that the relationship is linear – *and you would be wrong.*

There are a few clues. If you subtract the observed values of y from the values calculated from the regression equation, you will see that there is a systematic pattern to the errors. If you plot the data and the regression line, this pattern is evident (Fig. 12.2).

In this case, if you compute the regression of the logarithms, you obtain $\ln y = 1{\cdot}999 \ln x - 0{\cdot}6914$, and the correlation coefficient is 1·00 (Fig. 12.3). Restoring the natural logarithms ($\ln y = m \ln x + \ln b \rightarrow y = bx^m$), this suggests that the true relationship is $y = x^2/2$, which indeed it is. The result of the test of significance, that it is extremely unlikely that the association of points arose by chance, must not be interpreted as confirmation that it is *linear*. The test of significance indicates that it is most unlikely that the association arose by chance, and indeed, the association did not arise by chance.

The point is that *any* gentle curve, and any short part of a curve, will give good linear correlation. You must do the thinking, because statistics does not think for you. You must consider the boundary conditions. Take another example.

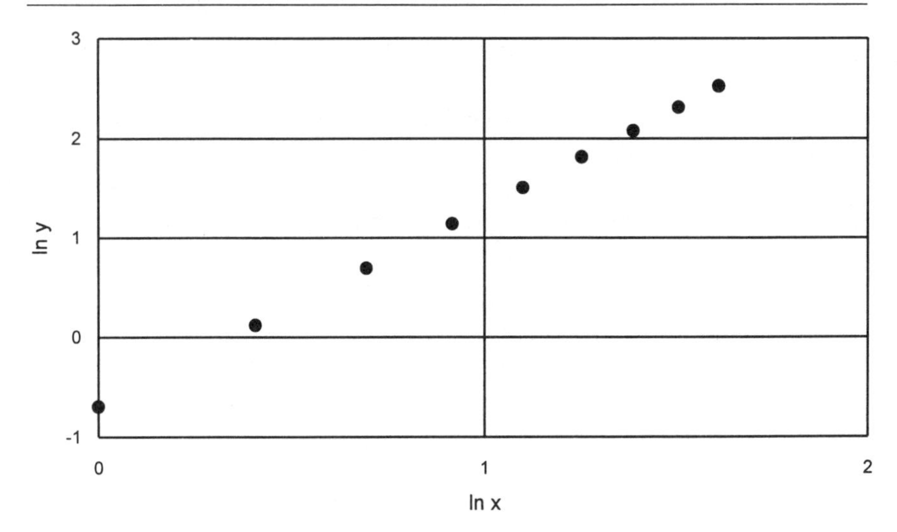

Figure 12.3 The same data plotted logarithm versus logarithm.
When the data listed in Table 12.1 are plotted as logarithms, the slope is evidently 2 or 0·5 (depending on which is the dependent variable)

You determine the acoustic or sonic velocity (V) and fractional porosity (f) in sandstones of different porosity so that you may determine the porosity by measuring the acoustic velocity. It is much easier to measure acoustic velocity than porosity in a borehole. You can measure both with considerable accuracy when you have a sample, so how would you set about the task of finding a formula relating the two? As mentioned on page 90, an association between a velocity, with dimensions LT^{-1} and porosity, which is dimensionless, must be in terms of a dimensionless velocity – in this case, the ratio of the measured velocity to the acoustic velocity when the porosity is zero. The boundary conditions are that when $f = 0$, $V = V_0$; and there is some maximum value of f, the value obtaining when the sediment accumulated into the stratigraphic record. This we labelled f_0 on page 90.

If you feed your data into a computer, and tell it to compute the linear correlation coefficients and so on, it will come up with a formula that appears quite satisfactory, with very small errors between prediction and measurement. Try plotting the solids proportion, $1–f$, against the ratio of the velocity observed to the velocity in the matrix, V/V_0 – no, the *logarithms* of those quantities – anticipating an equation of the dimensionless form

$$1 - f = b\,(V/V_0)^m$$

where V_0 is the acoustic velocity when the porosity is zero. Linear regression of $\ln(1-f)$ and $\ln(V/V_0)$ or $\log(1-f)$ and $\log(V/V_0)$ may suggest the values of b and m; but the warnings above apply here too – *any* gentle curve may appear to be linear with statistical significance.

What about the fluid in the pores? The short answer is that the acoustic velocity in quartz is about four times that in liquids, so if the wavelength is very short, the path of first arrival will be through pore fluid only over very short sections close to the points of contact between grains in uncemented sandstones. And because the path will be in general far from tangential to the grain, very little of it will be through the pore fluid. If the sandstone is cemented, virtually none of the path will be through fluid.

Table 12.2 Data for solids proportion and sonic velocity correlation.

$(1-f)$	V/V_0
0·824	0·701
0·819	0·683
0·827	0·649
0·785	0·642
0·783	0·626
0·779	0·625
0·780	0·610
0·770	0·609
0·765	0·606
0·761	0·606
0·761	0·595
0·755	0·565
0·758	0·560

Table 12.2 gives 13 measured values of the solid proportion, $1-f$, and the dimensionless velocity, V/V_0, in sandstone assuming a matrix velocity, V_0, of $5550\,\mathrm{m\,s^{-1}}$. Linear regression analysis gives the formula

$$(1-f) = 0{\cdot}57\ V/V_0 + 0{\cdot}43,$$

$$f = 0{\cdot}57 - 0{\cdot}57\ V/V_0$$

with a correlation coefficient of $r = 0{\cdot}90$ with 11 degrees of freedom, which is highly significant. This seems to be entirely satisfactory. However, linear regression analysis of the logarithms of the quantities reveals that the natural relationship cannot be linear because the slope is much nearer 0·5 than 1. The correlation coefficient is also 0·90.

Linear regression analysis of the logarithms leads to the following equation:

$$(1-f) = 0{\cdot}97\ (V/V_0)^{0{\cdot}45}$$

or

$$f = 1 - 0{\cdot}97\ (V/V_0)^{0{\cdot}45}.$$

Now, one boundary condition is that when the porosity is zero and $1-f = 1$, the dimensionless velocity $V/V_0 = 1$. The coefficient b must therefore also equal 1, not 0·97. If we constrain the line so that it passes through the points ln(1, 1) and the mean values of $\ln(1 - f)$ and $\ln(V/V_0)$ the slope is 0·52 and a good practical formula turns out to be

$$f = 1 - (V/V_0)^{0 \cdot 5}.$$

This satisfies the boundary conditions, and it only remains to determine the best value of V_0 to use. This is a very different formula from the one that describes seismic P waves on page 101, probably because the wavelengths of seismic P waves are very much longer than those of an acoustic borehole logging tool. The P wave generated by an earthquake has a frequency of about 1 Hz and a velocity of about 5 km s^{-1}, so its wavelength is about 5 km. The P wave generated in seismic exploration has a frequency of about 50 Hz and a velocity of about 3 km s^{-1}, so its wavelength is of the order of 60 m. The acoustic log signal (20 kHz) has character over less than 1 m and so its wavelength is very much less than 1 m – in fact, about 15 cm.

For a painless but instructive introduction to statistical methods, even if he predated the pocket calculator days, see Moroney (1956). See also Rosen (1992: 178ff.) for a good summary of the mathematical statistical techniques.

APPENDIX

Pronunciation

I take the liberty of suggesting the pronunciation of a few words and names that sometimes give difficulty. In doing this, I am aware, as you must be aware, of the advice given by H. W. Fowler in his *A dictionary of modern English usage* (Oxford University Press, 1926): "Pronounce as your neighbours do, not better". And, on the pronunciation of French words; "All that is necessary is a polite acknowledgement of indebtedness to the French language indicated by some approach in some part of the word to the foreign sound . . .".

In the SI units with prefixes, the emphasis is on the first syllable: nan'ometre (**nann**'ometer), mill'imetre, cent'imetre, and so also, logically, kil'ometre (**kill**'ometer), like kil'ogram. The *tonne* rhymes with *con* (perhaps not in North America!).

Of the French scientists, Bouguer is pronounced *boo-gair* with a hard g; Fermat, *fair-ma*; Fresnel, *fray-nel*; Poiseuille, *pwa-seÿy* (very roughly); Poisson, *pwa-son* rhyming with *on*, roughly.

The Dutch scientist Huygens (sometimes spelled Huyghens) is pronounced *high'gens*, roughly, with a hard g.

The Greek alphabet

Those characters that can be confused are not used as symbols.

α	alpha	Α	ν	nu	Ν
β	beta	Β	ο	omicron	Ο
γ	gamma	Γ	π	pi	Π
δ	delta	Δ	ρ	rho	Π
ε	epsilon	Ε	σ	sigma	Σ
ζ	zeta	Ζ	τ	tau	Τ
η	eta	Η	υ	upsilon	Υ
θ	theta	Θ	φ, φ	phi	Φ
ι	iota	Ι	χ	chi	Ξ
κ	kappa	Κ	ψ	psi	Ψ
λ	lambda	Λ	ω	omega	Ω
μ	mu	Μ			

χ is pronounced "keye", rhyming with "eye", not "chi" as in "China".

Answer to problem on buoyancy (p. 23)

When you put the dinghy in the water, the water level in the lake rises. You get into the dinghy and the water level rises. It rises when you take the stone on board, and again when you take the wood on board.

When you throw the stone overboard, the water level falls, but not by as much as it rose when the stone was put in the dinghy. In the dinghy, the stone displaces its *weight* of water: in the water, it sinks and displaces its *volume* of water. It displaces more water when it is in the dinghy than it does when it is thrown overboard.

There is no change when you throw the wood overboard. The wood floats, so it displaces its *weight* of water in the pond as well as in the boat.

Miscellaneous quantities

Angstrom (Ångström)	Å	100×10^{-12} m
Planck constant	h	$(662{\cdot}56 \pm 0{\cdot}05) \times 10^{-36}$ Js
gas constant	R	$(8{\cdot}314\,3 \pm 0{\cdot}0012)$ JK^{-1}mol^{-1}
speed of light in vacuum	c_o	$(299{\cdot}792\,5 \pm 0{\cdot}0003) \times 10^{6}$ ms^{-1}
Earth's equatorial radius	a	6378·14 km
Earth's polar radius	b	6356·77 km
Earth's mean radius, $(2a+b)/3$	R_E	6371 km
Earth's flattening, $f = (a-b)/a$	f	1/298·258
Earth's mean mass density		5517 kgm^{-3} (~3000 to ~13000 kgm^{-3})
Earth's mass	M_E	$5{\cdot}977 \times 10^{24}$ kg
Earth's mass/Moon's mass		81·3
Sun's mass/Earth's mass		333×10^{3}
mean distance Earth to Moon		$384{\cdot}4 \times 10^{3}$ km
mean distance Earth to Sun		149×10^{6} km
gravitational constant	G	$(66{\cdot}70 \pm 0{\cdot}05) \times 10^{-12}$ Nm^{2}kg^{-2}
geocentric gravitational constant	GM	$398{\cdot}603 \times 10^{12}$ m^{3}s^{-2}
g at sea level in latitude 40°		9·80616 ms^{-2}
g at Equator		9·780318 ms^{-2}
g at Poles		9·832177 ms^{-2}
period of Earth's rotation		86164·1 s (23h 56min 4·1 s)
magnetic permeability of vacuum	μ_o	$400p \times 10^{-9}$ kg m s^{-2} A^{-2}
velocity of sound, air, 20°C		343 ms^{-1}
velocity of sound, water, 20°C		1478 ms^{-1}

NOTES

1. The centripetal force required to keep a body in circular orbit is mV^2/r, as we found in the dimensional analysis on page 10. The velocity of the body is $2\pi r/T$ where T is the period, the time taken for one orbit. So,

   $F_c = ma = mV^2/r = 4\pi^2 r/T^2$, and

   $a = 4\pi^2 r/T^2$.

   Kepler's third law states that the square of the period is proportional to the cube of the mean distance, so

   $a \propto 1/r^2$.

   Similarly, given the inverse square law, you can infer Kepler's third law of planetary motion, that $r^3 \propto T^2$.

2. If you take the Earth's radius as the unit of distance, then, as stated in the text, the attraction of the Moon is roughly proportional to $1/60^2$ at the Earth's centre, $1/59^2$ at the sublunar point, and $1/61^2$ at the antipode. If x is the number of Earth's radii between the centre of the Earth and the centre of the Moon, the force at the sublunar point is proportional to $1/(x-1)^2$; at the centre of the Earth, to $1/x^2$; and at the antipode, to $1/(x+1)^2$. Two approximations are used. $(x-1)^2 = x^2-2x+1$ and since x is very large relative to 1, x^2-2x+1 is very nearly equal to x^2-2x which is equal to $x^2[1-(2/x)]$. The second approximation is that $(1-a)^{-1}$ is very nearly equal to $(1+a)$. This will be demonstrated below. So, the difference between the force at the sublunar point and the force at the centre of the Earth is proportional to

   $$\frac{1}{(x-1)^2} - \frac{1}{x^2} \approx \frac{[1+(2/x)]}{x^2} - \frac{1}{x^2} = \frac{2}{x^3}.$$

   Similarly, the difference between force at the antipode and that at the Earth's centre is proportional to $-2/x^3$.

   The second approximation can be demonstrated as follows:

   Set up an equation: $1/(1-a) = 1 + b$. So,

   $(1+b)(1-a) = 1 + b - a - ab = 1$.

   If a and b are small relative to 1, the product ab can be neglected, and a is very nearly equal to b, and $1/(1-a)$ is very nearly equal to $1+a$.

   The approximation is also the first two terms of the binomial expansion $(1-a)^{-1} = 1 + a + a^2 + \ldots$ for $a^2 < 1$, in this case, very much less than one.

   Returning to the quantities in the text, the Earth's radius R_E and the distance to the moon d_M,

   $$\frac{1}{(d_M - R_E)^2} \approx \frac{1}{d_M^2\,[1-(2R_E/d_M)]} \approx \frac{1+(2R_E/d_M)}{d_M^2}$$

   and the difference is

   $$\frac{1+(2R_E/d_M)}{d_M^2} - \frac{1}{d_M^2} \approx \frac{2R_E}{d_M^3}.$$

3. Let the velocities of two parallel wave-trains travelling in the same direction in deep water differ by dV, and their wavelengths by $d\lambda$ – the faster waves having the longer wavelength. Each time the

faster, longer wave passes the shorter and slower by $d\lambda$, the coincidence of waves will fall back by one wavelength in the time $d\lambda/dV$. So the group velocity will be $V - \lambda(dV/d\lambda)$. In deep water, $V = (g\lambda/2\pi)^{1/2}$, and $\lambda(dV/d\lambda) = (\lambda/2)(g/2\pi\lambda)^{1/2} = \frac{1}{2}(g\lambda/2\pi)^{1/2} = V/2$. See, for example, Tricker (1984) for a fuller discussion of water waves, and John (1984) of their motion.

4. The meaning of the term *degrees of freedom* is only superficially simple. In the matter of correlation, *any* two points will lie on a straight line. Even two points on the circumference of a circle will lie on a straight line, but not three. Two points provide no information about the relationship between the two. If you have ten paired measurements and you wish to assess their degree of correlation, any test of significance must be carried out for (10–2) degrees of freedom. More generally, if there are *n* pairs, there are (*n*–2) degrees of freedom in regression analysis.
If you make 10 independent measurements of, for example, the ages and heights of some students in a school, then there are 10 degrees of freedom. But if you continue the measurements until you have 1000, and record the numbers within each age group, e.g. 5, 6, 7 and 8, and within each 10 cm of height from 1·2 to 1·5 m, you might draw up a table (called a contingency table) like this:

Height (m)	Age (years)			
	5	6	7	8
1·401–1·500	–	–	–	–
1·301–1·400	–	–	–	–
1·201–1·300	–	–	–	–
1·101–1·200	–	–	–	–

The frequencies in each row are not all independent because when three are known and we know the total, the fourth is determined and a degree of freedom is lost. Likewise, the frequencies in each column are not all independent, and a degree of freedom is lost. So the degrees of freedom for the table are not 4×4 but 3×3, or, more generally, where *r* is the number of rows, and *c*, the columns, $(r-1)(c-1)$. Moroney (1956) is not very enlightening on this subject. When in doubt, the best advice, as always, is to ask someone who knows these things.

As regards the assessment of significance in linear regression analysis, the purpose is to establish if such a distribution of points could reasonably have arisen by chance. The result is never that it could not have arisen by chance, but a probability, *P*, that it arose by chance. The usual values chosen as levels of significance are five per cent, one per cent and 0·1 per cent. Perhaps the easiest to use is Student's *t*, or the *t*-test, for the significance of the regression coefficient, *r*. This was used in the examples. For this, you calculate

$$t = \frac{(r\sqrt{N-2})}{(\sqrt{1-r^2})}$$

where *N* is the number of pairs in the analysis and *N*–2 are the degrees of freedom, and enter this value of *t* into the tables of *t* with *N*–2 degrees of freedom. From the table we assess the probability *P* of having *t* as large *or larger* than some critical number. If *r* is negative, this should be ignored in this test. Most calculators (and spreadsheets, no doubt) include the statistical operations, and some include the significance tests. For methods of assessing the significance of a regression analysis, see Moroney (1956: 311*f*).

It is worth repeating the warning that if a test of significance of a linear regression analysis indicates that the probability *P* is very small of obtaining such a result by chance, it does not necessarily confirm the analysis. It is very important to establish linearity by plotting the logarithms of the data, or by computing the regression coefficient of the logarithms. As mentioned in the text, *any* gentle curve or short section of a curve, will appear to be significant.

REFERENCES

Allum, J. A. E. 1966. *Photogeology and regional mapping*. Oxford: Pergamon.

Bagnold, R. A. 1941. *The physics of blown sand and desert dunes*. London: Methuen.

Bates, R. L. & J. A. Jackson (eds) 1987. *Glossary of geology*, 3rd edn. Alexandria, Virginia: American Geological Institute.

Bleany, B. 1984. Magnetism. In *The new Encyclopædia Britannica*, 15th edn, vol. 11, 309–28. Chicago: Encyclopædia Britannica.

Brinkworth, B. J. 1968. *An introduction to experimentation*. London: English Universities Press.

Buckingham, E. 1914. On physically similar systems: illustrations of the use of dimensional equations. *Physical Review*, 2nd series, **4**, 345–76.

— 1921. Notes on the methods of dimensions. *London Edinburgh Dublin Philosophical Magazine*, 6th series, **42**, 696–719.

Carey, S. W. 1954. The rheid concept in geotectonics. *Journal of the Geological Society of Australia* **1** (for 1953), 67–117.

Clark, S. P. 1966. *Handbook of physical constants*, revised edn. Geological Society of America, Memoir 97.

Chapman, R. E. 1979. Mechanics of unlubricated sliding. *Bulletin of the Geological Society of America* **90**, 19–28.

— 1981. *Geology and water: an introduction to fluid mechanics for geologists*. The Hague: Martinus Nijhoff/Junk.

Faure, G. 1986. *Principles of isotope geology*, 2nd edn. New York: John Wiley.

Feynman, R. P., R. B. Leighton, M. Sands 1963–5. *The Feynman Lectures on Physics*, 3 volumes. Reading, Mass.: Addison-Wesley. [Vol. 1, 1963; vol. 2, 1964; vol. 3, 1965]

Hubbert, M. K. 1937. Theory of scale models as applied to the study of geologic structures. *Geological Society of America, Bulletin* **48**, 1459–520.

— 1940. The theory of ground-water motion. *Journal of Geology* **48**, 785–944.

— 1951. Mechanical basis for certain familiar geologic structures. *Geological Society of America, Bulletin* **62**, 355–72.

— 1969. *The theory of ground-water motion and related papers*. New York: Hafner.

— & W. W. Rubey 1959. Role of fluid pressure in mechanics of overthrust faulting, I. Mechanics of fluid-filled porous solids and its application to overthrust faulting. *Geological Society of America, Bulletin* **70**(2), 115–66.

Jaeger, J. C., & N. G. W. Cook 1979. *Fundamentals of rock mechanics*, 3rd edn. London: Chapman & Hall. [But ignore Chapter 8 on fluid flow, which is erroneous.]

John, W. 1984. Wave motion. In *The new Encyclopædia Britannica*, 15th edn, vol. 19, 665–73. Chicago: Encyclopædia Britannica.

Kehle, R. O. 1970. Analysis of gravity sliding and orogenic translation. *Geological Society of America, Bulletin* **81**(6), 1641–64.

Mohr, O. C. 1882. Über die Darstellung des Spannungszustandes und des Deformationszustandes eines Körperelementes und über die Anwendung derselben in der Festigkeitslehre. *Der Civilingenieur* **28**, 113–56.

Moroney, M. J. 1956. *Facts from figures*, 3rd edn. London: Penguin.

Pankhurst, R. C. 1964. *Dimensional analysis and scale factors*. London: Chapman & Hall.

Ramberg, H. 1981. *Gravity, deformation, and the Earth's crust: in theory, experiments, and geological application*, 3rd edn. London: Academic Press.

Richter, C. F. 1984. Earthquakes. In *The new Encyclopædia Britannica*, 15th edn, vol. 6, 68–73.

Chicago: Encyclopædia Britannica.

Rosen, D. 1992. *Mathematics recovered for the natural and medical sciences.* London: Chapman & Hall.

Smoluchowski, M. S. 1909. Some remarks on the mechanics of overthrusts. *Geological Magazine*, new series, Decade V, **6**, 204–205.

Steiger, R. H., & E. Jäger 1977. Subcommission on geochronology: convention on the use of decay constants in geo- and cosmochemistry. *Earth and Planetary Science Letters* **36**, 359–62.

Telford, W. M., L. P. Geldart, R. E. Sheriff 1990. *Applied geophysics*, 2nd edn. Cambridge: Cambridge University Press.

Terzaghi, K. 1943. *Theoretical soil mechanics.* London: Chapman & Hall; New York: John Wiley.

— 1950. Mechanism of landslides. In *Application of geology to engineering practice* (Berkey vol.), S. Paige (Chairman), 83–124. Boulder, Colorado: Geological Society of America.

Tricker, R. A. R. 1984. Water waves. In *The new Encyclopædia Britannica*, 15th edn, vol. 19, 654–60. Chicago: Encyclopædia Britannica.

Turner, J. S. 1973. *Buoyancy effects in fluids.* Cambridge: Cambridge University Press.

Turcotte, D. L. & G. Schubert 1982. *Geodynamics: applications of continuum physics to geological problems.* New York: John Wiley. [Beware of Darcy's Law, which is in the form valid for horizontal flow only: and do not refer to Muskat's *The flow of homogeneous fluids through porous media*, which is probably where this treatment of Darcy's law came from.]

FURTHER READING

Bascom, W. 1980. *Waves and beaches: the dynamics of the ocean surface*, 2nd edn. New York: Anchor Press/Doubleday.

Born, M. 1949. *Natural philosophy of cause and chance* (Waynefleet lectures). Oxford: Oxford University Press.

Darcy, H. 1856. *Les fontaines publiques de la ville de Dijon.* Paris: Victor Dalmont. [The appendix that contains the results of the experiments has been reprinted in Hubbert (1969).]

Dugdale, D. S. 1984. Elasticity. In *The new Encyclopædia Britannica*, 15th edn, vol. 6, 519–22. Chicago: Encyclopædia Britannica.

Elwyn, A. J. 1984. Neutron. In *The new Encyclopædia Britannica*, 15th edn, vol. 12, 1070–76. Chicago: Encyclopædia Britannica.

The new Encyclopædia Britannica, 15th edn, 1984. Chicago: Encyclopædia Britannica.

Goetz, P. W., & M. Sutton (eds) 1984. *The new Encyclopædia Britannica*, 15th edn. Chicago: Encyclopædia Britannica.

Graham, L. 1984. Heat. In *The new Encyclopædia Britannica*, 15th edn, vol. 18: 700–6. Chicago: Encyclopædia Britannica.

Holmes, A., 1965. *Principles of physical geology*, 2nd edn. London: Nelson.

Hubbert, M. K. 1945. The strength of the Earth. *American Association of Petroleum Geologists, Bulletin* **29**, 1630–53.

— 1963. Are we retrogressing in science? *Geological Society of America,* Bulletin **74**, 365–78.

— 1972. *Structural geology.* New York: Hafner. [Collected papers]

Keenan, J. H., G. N. Hatsopoulos, E. P. Gyftopoulos 1984. Principles of thermodynamics. In *The New Encyclopædia Britannica*, 15th edn. Chicago: Encyclopædia Britannica **18**, 290–315.

Lindsay, R. B. 1984. Sound. In *The new Encyclopædia Britannica*, 15th edn, vol. 15, 19–34. Chicago: Encyclopædia Britannica.

Metric Conversion Board 1973. *Metric practice.* Canberra: Australian Government Publishing Service.

Phillips, M. 1984. Electromagnetic radiation. In *The new Encyclopædia Britannica*, 15th edn, vol. 6, 645–65. Chicago: Encyclopædia Britannica.

Raitt, G. 1987. *Heat and temperature.* Cambridge: Cambridge University Press.

Ramberg, H. 1981. *Gravity, deformation, and the Earth's crust: in theory, experiments, and geological application*, 2nd edn. London: Academic Press.

Ramsay, J. G. 1967. *Folding and fracturing of rocks.* New York: McGraw-Hill.

Sorby, H. C. 1908. On the application of quantitative methods to the study of the structure and history of rocks. *Geological Society of London, Quarterly Journal* **64**, 171–233.

Stoner, J. O. & E. B. Wilson 1984. Spectroscopy, principles of. In *The new Encyclopædia Britannica*, 15th edn, vol. 17, 455–76. Chicago: Encyclopædia Britannica.

Trigg, G. L. & S. A. Goudsmit 1984. Atomic structure. In *The new Encyclopædia Britannica*, 15th edn, vol. 2, 330–43. Chicago: Encyclopædia Britannica.

Tuma, J. J. 1976. *Handbook of physical calculations.* New York: McGraw-Hill.

Ziman, J. M. 1984. Electron. In *The new Encyclopædia Britannica*, 15th edn, vol. 6, 665–72. Chicago: Encyclopædia Britannica.

INDEX